ENGINEERING PRACTICES FOR AGRICULTURAL PRODUCTION AND WATER CONSERVATION

An Interdisciplinary Approach

Innovations in Agricultural and Biological Engineering

ENGINEERING PRACTICES FOR AGRICULTURAL PRODUCTION AND WATER CONSERVATION

An Interdisciplinary Approach

Edited by
Megh R. Goyal, PhD, PE
R. K. Sivanappan, PhD

Apple Academic Press Inc.
3333 Mistwell Crescent
Oakville, ON L6L 0A2 Canada

Apple Academic Press Inc.
9 Spinnaker Way
Waretown, NJ 08758 USA

©2017 by Apple Academic Press, Inc.

First issued in paperback 2021

Exclusive worldwide distribution by CRC Press, a member of Taylor & Francis Group
No claim to original U.S. Government works

ISBN 13: 978-1-77-463045-7 (pbk)
ISBN 13: 978-1-77-188451-8 (hbk)

Library and Archives Canada Cataloguing in Publication

Engineering practices for agricultural production and water conservation: an interdisciplinary approach / edited by Megh R. Goyal, PhD, PE, R.K. Sivanappan, PhD.

(Innovations in agricultural and biological engineering)
Includes bibliographical references and index.
Issued in print and electronic formats.
ISBN 978-1-77188-451-8 (hardcover).--ISBN 978-1-77188-452-5 (pdf)

1. Water conservation. 2. Soil conservation. 3. Agricultural engineering. I. Goyal, Megh Raj, editor II. Sivanappan, R. K., author, editor III. Series: Innovations in agricultural and biological engineering

TD388.E55 2016   631   C2016-907099-9 C2016-907100-6

Library of Congress Cataloging-in-Publication Data

Names: Goyal, Megh Raj, editor. | Sivanappan, R. K., editor.
Title: Engineering practices for agricultural production and water conservation : an interdisciplinary approach / editors: Megh R. Goyal, R. K. Sivanappan.
Description: Waretown, NJ : Apple Academic Press, 2017. | Includes bibliographical references and index.
Identifiers: LCCN 2016048790 (print) | LCCN 2016050098 (ebook) | ISBN 9781771884518 (alk. paper) | ISBN 9781771884525 (ebook)
Subjects: LCSH: Water conservation. | Soil conservation. | Agricultural engineering.
Classification: LCC TD388 .E55 2017 (print) | LCC TD388 (ebook) | DDC 631--dc23
LC record available at https://lccn.loc.gov/2016048790

Apple Academic Press also publishes its books in a variety of electronic formats. Some content that appears in print may not be available in electronic format. For information about Apple Academic Press products, visit our website at **www.appleacademicpress.com** and the CRC Press website at **www.crcpress.com**

CONTENTS

LIST OF CONTRIBUTORS

Brahamdutt Arya, MSc
Assistant Professor, Department of Chemistry, Pt. Neki Ram College, Rohtak – 124001, Haryana, India

Vedpriya Arya, PhD
Assistant Professor; Scientist E, Patanjali Herbal Research Department, Patanjali Yogpeeth, Phase I, Haridwar, Uttrakhand; Mobile: +91-7060472471, E-mail: ved.nano2008@gmail.com

Bavita Asthir, PhD
Assistant Professor, Department of Biochemistry, Punjab Agricultural University, Ludhiana 141004, Punjab, India

Nitin Dabral, BTech
Research Scholar MTech, Department of Agricultural Engineering, College of Technology, G.B.P.U.A.T., Pantnagar, India. E-mail: nitin11dabral@gmail.com

R. Doraiswamy, PhD
Institutional Expert, Water Resources and Development, Bangalore – 560052, India. E-mail: doraiswamyram@gmail.com

Megh R. Goyal, PhD, PE
Retired Professor in Agricultural and Biomedical Engineering, University of Puerto Rico – Mayaguez Campus; and Senior Technical Editor-in-Chief in Agriculture Sciences and Biomedical Engineering, Apple Academic Press Inc., PO Box 86, Rincon – PR – 00677 – USA. E-mail: goyalmegh@gmail.com

Lijuan Jiang, PhD
Professor, Central South University of Forestry and Technology, 498 South Shaoshan Rd., Changsha, Hunan 410004, China

Manbir Kaur, MSc
Department of Biotechnology, Guru Nanak Girls College, Ludhiana, Punjab, India

M. K. Khaishagi, PhD
Irrigation Engineer, Water Resources and Irrigation Engineering, Hyderabad, India – 500034. E-mail: mkkhaishagi@yahoo.com

Brajesh Kumar, PhD
Professor, Department of Chemistry, TATA College, Kolhan University, Chaibasa – 833202, Jharkhand, India. Current address: Centro de Nanociencia y Nanotecnologia, Universidad de las Fuerzas Armadas -ESPE, Av. Gral. Rumiñahui s/n, Sangolqui, P.O. Box 171-5-231B, Ecuador; E-mail: krmbraj@gmail.com; bkumar@espe.edu.ec

Mayank Kumar, BTech
Research Scholar MTech, Indian Institute of Technology at Bombay Room No. B-003C, Hostel No. 12, IIT Bombay, Powai, Mumbai – 400076, India. E-mail: mayank1992kumar@gmail.com

Changzhu Li, PhD
Professor, Hunan Academy of Forestry, 658 South Shaoshan Rd., Changsha, Hunan – 410004, China

Peiwang Li, M.S.
Associate Professor, Hunan Academy of Forestry, 658 South Shaoshan Rd., Changsha, Hunan 410004, China

Shrikant Daji Limaye, PhD
Director, Ground Water Consultant for at Several Industries & Agri. Banks; UNESCO – IUGS – IGCP Project 523 "GROWNET": Ground Water Institute, Pune – India.
http://www.igcp-grownet.org; E-mail: limaye@vsnl.com

Laszlo Mago, PhD
Professor, SzentIstván University, Faculty of Mechanical Engineering, Gödöllő, Hungary

J. A. Moses, PhD
Assistant Professor, Indian Institute of Crop Processing Technology (IICPT), MoFPI, Govt. of India, Pudukottai Road, Thanjavur, Tamil Nadu – 613005, India. E-mail: moses.ja@iicpt.edu.in

G. R. Ramakrishna Murthy, PhD
Principal Scientist, Education Systems Management Division, National Academy of Agricultural Research Management (NAARM), Rajendranagar, Hyderabad – 500030, India. E-mail: murthy@naarm.ernet.in

K. Palanisami, PhD
Emeritus Scientist, International Water Management Institute (IWMI), ICRISAT campus, Patanchereu – 502324, Hyderabad, India. Mobile: +91-9000686853; E-mail: k.palanisami@cgiar.org

Thaneswer Patel, PhD
Assistant Professor, Department of Agricultural Engineering, North Eastern Regional Institute of Science and Technology (NERIST), Nirjuli (Itanagar), Arunachal Pradesh–791109, India. Mobile: +91-9436228996, E-mail: thaneswer@gmail.com; athaneswer@iitg.ernet.in

P. K. Pranav, PhD
Assistant Professor, Department of Agricultural Engineering, North Eastern Regional Institute of Science and Technology (NERIST), Nirjuli (Itanagar) – 791109, Arunachal Pradesh, India. E-mail: Pkjha78@gmail.com

Dušan Radojičić, PhD
Professor, University of Belgrade – Faculty of Agriculture, Department of Agricultural Engineering, Nemanjina 6, 11080 Belgrade, Serbia

Y. K. Rao, MTech
Senior Agricultural Engineer, Southern Region Farm Machinery Training and Testing Institute, Tractor Nagar, Garladinne, Anantapur (Dist.) – 515731, Andhra Pradesh, India. E-mail: ykrao1969@yahoo.in

K. M. Singh, PhD
Director, Extension Education and Chair, Department of Agricultural Economics, Rajendra Agricultural University, Bihar, Pusa-Samastipur, India – 848125. Mobile: +91-9431060157; E-mail: krishna_singh2000in@yahoo.com; m.krishna.singh@gmail.com

K. P. Singh, PhD
Scientist, Directorate of Maize Research, Pusa Campus, New Delhi – 110012, India

Rajni Singh, PhD
Additional Director, Amity Institute of Microbial Biotechnology, Room No. 307, J-3 Block, Amity University Uttar Pradesh, Sector-125, Noida, U.P., India. Tel.: +91-120-4392900; E-mail: rajni_vishal@yahoo.com; rsingh3@amity.edu

S. P. Singh, PhD
Scientist, National Physical Laboratory, Dr. K.S. Krishnan Marg, New Delhi – 110012, India

R. K. Sivanappan, PhD
Former Professor and Dean, College of Agricultural Engineering & Technology, Tamil Nadu Agricultural University (TNAU), Coimbatore. Current mailing address: Consultant, 14, Bharathi Park, 4th Cross Road, Coimbatore, Tamilnadu – 641043, India, E-mail: sivanappanrk@hotmail.com

Prem Prakash Srivastav, PhD
Agricultural and Food Engineering Department, Indian Institute of Technology, Kharagpur 721302, West Bengal, India

Shikha Srivastava, PhD
Department of Botany, Deen Dyal Upadhyay Gorakhpur University, Gorakhpur – 273009, Uttar Pradesh, India

Youping Sun, PhD
Research Associate, Texas A&M Agri Life Research Center at El Paso, 1380 A&M Circle, El Paso, TX 79927, USA. Mobile: +001-8646339385; E-mail: youping.sun@ag.tamu.edu

Goran Topisirovic, PhD
Full Professor & Chair of Agricultural Engineering, Faculty of Agriculture, Institute of Agricultural Engineering, University of Belgrade, Nemanjina 6, 11080 Belgrade-Zemn, Serbia. Skype: gogibog, Tel.: +381 112199621; Mobile: +381 692417125 E-mail: gogi@agrif.bg.ac.rs

Vinod Kumar Tripathi, PhD
Assistant Professor, Centre for Water Engineering & Management, Central University of Jharkhand, Ranchi – Pin code, India. E-mail: tripathiwtcer@gmail.com

Deepak Kumar Verma, PhD
Research Scholar and DST INSPIRE Fellow, Agricultural and Food Engineering Department, Indian Institute of Technology, Kharagpur – 721302, West Bengal, India. Tel: +91–3222–281673. Mobile: +91-7407170260, +91-9335993005. Fax: +91-3222-282224; Mailing address: Room No. NW-4, B.C. Roy Hall of Residence, Indian Institute of Technology, Kharagpur, West Bengal – 721302, India. E-mail: deepak.verma@agfe.iitkgp.ernet.in; rajadkv@rediffmail.com

P. Vithu, BTech (FPE)
Research Scholar MTech, Indian Institute of Crop Processing Technology (IICPT), MoFPI, Govt. of India, Pudukottai Road, Thanjavur, Tamil Nadu – 613005, India

Zhihong Xiao, PhD
Professor, Hunan Academy of Forestry, 658 South Shaoshan Rd., Changsha, Hunan 410004, China

Liangbo Zhang, PhD
Associate Professor, Hunan Academy of Forestry, 658 South Shaoshan Rd., Changsha, Hunan 410004, China

Ivan Zlatanović, PhD
Professor, University of Belgrade – Faculty of Agriculture, Department of Agricultural Engineering, Nemanjina 6, 11080 Belgrade, Serbia

LIST OF ABBREVIATIONS

6-BA, N^6	benzyladenine
ABA	abscisic acid
AFM	atomic-force microscopy
Ag(I)	silver iodide
AICTE	All India Council for Technical Education
APS	advanced photon source
BIS	Bureau of Indian Standards
BOD	biochemical oxygen demand
C_{60}	Carbon 60
CII	Confederation of Indian Industries
COD	chemical oxygen demand
CRIDA	Central Research Institute for Dryland Agriculture
CSIR	Council for Scientific & Industrial Research
CV	coefficient of variation
D	diameter of ground wheel
DAP	diammonium phosphate
DBT	Department of Biotechnology
df	degrees of freedom
DKW	Driver and Kuniyuki Walnut Medium
DNA	deoxyribonucleic acid
DO	dissolved oxygen
DRDO	Defense Research & Development Organization
DSE	German Foundation for International Development
DST	Department of Science & Technology
DTT	dithiothreitol
F	f-test statistic
FICCI	Federation of Indian Chambers of Commerce and Industries
GA_3	gibberellic acid
GBPIHED	G.B. Pant Institute of Himalayan Environment & Development
GOI	Government of India

HRDI	Herbal Research and Development Institute
IAA	indole-3-acetic acid
IBA	3-indole butyric acid
ICAR	Indian Council of Agricultural Research
ICIMOD	International Center for Integrated Mountain Development
ICMR	Indian Council of Medical Research
K	potash
MADPs	medicinal aromatic and dye plants
MAPPA	Medicinal and Aromatic Plants Program in Asia
MAPs	medicinal and aromatic plants
MEs	micro-emulsions
MoEF	Ministry of Environment & Forest
MOP	muriate of potash
MS	mean square
MS	Murashige-Skoog
MSNs	meso-porous silica nano-particles
N	nitrogen
NAA	α-naphthaleneacetic acid
NABARD	National Bank for Agriculture and Rural Development
NEC	Nippon Electric Company
NEs	nano-emulsions
NGO	non-government organization
NLCs	nano-structure lipid carriers
NMPB	National Medicinal Plants Board
P	phosphorus
PC&M	Production-To-Consumption & Marketing
PEG	poly-ethoxy glycol
PGRs	plant growth regulators
PNs	polymeric nano-particles
PVP	polyvinylpyrrolidone
Q	nutrient requirement
RNA	ribonucleic acid
SD	standard deviation
SEI	Stockholm Environment Institute
SLNs	solid lipid nano-particles
SS	sum of squares

TBHQ	tert-butylhydroquinone
TDS	total dissolved solids
TiO_2	titanium dioxide
TN	Tamil Nadu
TNAU	Tamil Nadu Agricultural University
US	United States
USG	urea super granule
USLI	ultra large scale integration
VSLI	very large scale integration
WHO	World Health Organization
WOCMAP	World Congress on Medicinal and Aromatic Plants for Human Welfare
WPM	woody plant medium
WTO	World Trade Organization
Zeatin	4-hydroxy-3-methyltrans-2-butenylaminopurine

LIST OF SYMBOLS

@	at the rate
/	per
%	percent
A	mL of total HCl used with phenolphthalein and methyl orange indicators
Avg.	average
b	width of coverage
C	climate, that is total rainfall, its distribution and evaporation characteristics
c	cup width
Conc.	concentration
D	drainage conditions
ET	evapotranspiration during the rainy season
et al.	and others
h	cup height
J_m	the maximum current
kg/ha	kilogram per hectare
km/h	kilometer per hour
mg/L	milligram per liter
n	number of splits fertilizer requirement
P	precipitation
P_{in}	the power of incident light ($MWcm^{-2}$)
Q	quality of irrigation water
R	surface runoff
r	net recharge
S	soil type
V_m	the maximum voltage
μm	micro meter
ρ	density of fertilizer
φ	angle of repose
Φ	diameter

PREFACE 1 BY MEGH R. GOYAL

Resource, ASABE, November/December 2015, 22(6), 20–21 introduces *"cubic farming™ that is a form of controlled environment agriculture in a commercial reality for the urban sector. It is referred to as vertical farming. It allows growth of fresh produce year round in locations that are challenging for conventional forms of agriculture, including arid urban areas and harsh northern climates—This technology might seem far out for some consumers, but is an insight into the future fresh food production."*

This new technology poses challenges to agricultural engineers to come up with appropriate solutions to vertical farming. Apple Academic Press Inc. published my first book *Management of Drip/Trickle or Micro Irrigation*, 10-volume set under book series Research Advances in Sustainable Micro Irrigation, in addition to other books in the focus areas of Agricultural and Biological Engineering. The mission of this book volume is to introduce the profession of agricultural and biological engineering. I cannot guarantee the information in this book series will be enough for all situations.

At the 49th annual meeting of the Indian Society of Agricultural Engineers at Punjab Agricultural University during February 22–25 of 2015, a group of ABEs convinced me that there is a dire need to publish book volumes on the focus areas of agricultural and biological engineering (ABE). This is how the idea was born on new book series titled *Innovations in Agricultural & Biological Engineering*.

The contributions by the cooperating authors to this book volume have been most valuable in the compilation. Their names are mentioned in each chapter and in the list of contributors. This book would not have been written without the valuable cooperation of these investigators, many of whom are renowned scientists who have worked in the field of ABE throughout their professional careers.

Dr. R. K. Sivanappan—Former Dean and Professor at Tamil Nadu Agricultural University (TNAU), Founding Director of Water Technology Institute at TNAU, author of more than 1000 professional articles and over 30 books—joins as coeditor of this book volume. He is a frequent contributor to my book series and a staunch supporter of my profession. I have inherited many ethical qualities of a successful educator from his clean testimony. His contribution to the contents and quality of this book has been invaluable.

I thank the editorial staff, Sandy Jones Sickels, Vice President, and Ashish Kumar, Publisher and President at Apple Academic Press, Inc., for making every effort to publish the book when the diminishing water resources are a major issue worldwide. Special thanks are due to the AAP Production Staff for the quality production of this book.

I request that the reader offer his constructive suggestions that may help to improve the next edition.

I express my deep admiration to my family for understanding and collaboration during the preparation of this book. As an educator, there is a piece of advice to one and all in the world: *"Permit that our almighty God, our Creator and excellent Teacher, irrigate the life with His Grace of rain trickle by trickle, because our life must continue trickling on... and Get married to your profession."*

—*Megh R. Goyal, PhD, PE*
Senior Editor-in-Chief

PREFACE 2 BY R. K. SIVANAPPAN

This volume on *Engineering Practices for Agricultural Production and Water Conservation: An Interdisciplinary Approach* is being published under the book series *Innovations in Agricultural and Biological Engineering*. Agricultural and Biological Engineering has been added recently to the Engineering faculty, and this knowledge and technology are essential for increasing food production for the growing population in the world. It is gratifying to note that food production in the world at present more than meets the requirements, but the food wastages and fair access to all are a major issues. However, food production will have to be increased to cater to the needs of the escalating population of the world in the coming years. This book was prepared to help achieve the goal by the application of the latest technologies in food production saving water and fertilizer, reducing wastages and by using the latest technology like nanotechnology in agriculture.

The book contains six parts namely: soil and water conservation and management; agricultural processing engineering; water quality and management; scope of emerging agricultural crops; renewable energy use in agriculture; and the applications of nanotechnology in agriculture. Dr. Megh R. Goyal, Senior Editor-in-Chief of this book, has contacted many experts on this subject throughout the world and received contributions to include in this book. The contributions are from India, Russia, and USA. We both are thankful to the contributing authors of each chapter who have shared their expertise in a nice way. These chapters cover the knowledge of soil and water conservation and management practices, methods of rain water harvesting, sustainable groundwater in hard rock areas, providing drainage for increasing crop production, precision application of fertilizer, agricultural processing to save food grains from wastages, potential of renewable sources from solid biomass, scope of medicinal and aromatic plants farming, and applications of nanotechnology in agriculture.

My special thanks are due to Dr. Megh R. Goyal, who has taken arduous job of editing the chapters from authors from various countries. This book will help governments, scientists, researchers, farmers and students all over the world. My thanks are also due to my wife, Mrs. S. Kannammal, our son, Dr. S. Kumar (Vice President at Asian Institute of Technology, Bangkok), and our daughter, Dr. S. Uma, for encouraging my involvement in publishing this book.

—R. K. Sivanappan, PhD
Co-editor

WARNING/DISCLAIMER

READ CAREFULLY

The goal of this book volume on *Engineering Practices for Agricultural Production and Water Conservation: An Interdisciplinary Approach* is to guide the world community on how to manage efficiently the technology available for different processes in food engineering. The reader must be aware that the dedication, commitment, honesty, and sincerity are most important factors in a dynamic manner for complete success. It is not a one-time reading of this compendium. Read and follow every time, it is needed.

The editors, the contributing authors, the publisher and the printer have made every effort to make this book as complete and as accurate as possible. However, there still may be grammatical errors or mistakes in the content or typography. Therefore, the contents in this book should be considered as a general guide and not a complete solution to address any specific situation in food engineering. For example, one type of food process technology does not fit all case studies in dairy engineering/science/technology.

The editors, the contributing authors, the publisher and the printer shall have neither liability nor responsibility to any person, any organization or entity with respect to any loss or damage caused, or alleged to have caused, directly or indirectly, by information or advice contained in this book. Therefore, the purchaser/reader must assume full responsibility for the use of the book or the information therein.

The mention of commercial brands and trade names is only for technical purposes. This does not mean that a particular product is endorsed over to another product or equipment not mentioned.

All weblinks that are mentioned in this book were active on June 30, 2015. The editors, the contributing authors, the publisher and the printing company shall have neither liability nor responsibility, if any of the weblinks is inactive at the time of reading of this book.

OTHER BOOKS ON AGRICULTURAL AND BIOLOGICAL ENGINEERING BY APPLE ACADEMIC PRESS, INC.

Management of Drip/Trickle or Micro Irrigation
Megh R. Goyal, PhD, PE, Senior Editor-in-Chief

Evapotranspiration: Principles and Applications for Water Management
Megh R. Goyal, PhD, PE, and Eric W. Harmsen, Editors

Book Series: Research Advances in Sustainable Micro Irrigation
Senior Editor-in-Chief: Megh R. Goyal, PhD, PE

Book Series: Innovations and Challenges in Micro Irrigation
Senior Editor-in-Chief: Megh R. Goyal, PhD, PE
> Volume 1: Principles and Management of Clogging in Micro Irrigation
> Volume 2: Sustainable Micro Irrigation Design Systems for Agricultural Crops: Methods and Practices
> Volume 3: Performance Evaluation of Micro Irrigation Management: Principles and Practices
> Volume 4: Potential of Solar Energy and Emerging Technologies in Sustainable Micro Irrigation
> Volume 5: Micro Irrigation Management: Technological Advances and Their Applications
> Volume 6: Micro Irrigation Engineering for Horticultural Crops: Policy Options, Scheduling, and Design
> Volume 7: Micro Irrigation Scheduling and Practices
> Volume 8: Engineering Interventions in Sustainable Trickle Irrigation: Water Requirements, Uniformity, Fertigation, and Crop Performance

Book Series: Innovations in Agricultural and Biological Engineering
Senior Editor-in-Chief: Megh R. Goyal, PhD, PE
- Modeling Methods and Practices in Soil and Water Engineering
- Food Engineering: Modeling, Emerging issues and Applications.
- Emerging Technologies in Agricultural Engineering
- Dairy Engineering: Advanced Technologies and their Applications
- Food Process Engineering: Emerging Trends in Research and Their Applications
- Soil and Water Engineering: Principles and Applications of Modeling
- Soil Salinity Management in Agriculture: Technological Advances and Applications
- Developing Technologies in Food Science: Status, Applications, and Challenges
- Engineering Practices for Agricultural Production and Water Conservation: An Interdisciplinary Approach
- Flood Assessment: Modeling and Parameterization
- Technological Interventions in the Processing of Fruits and Vegetables
- Technological Interventions in Management of Irrigated Agriculture
- Food Technology: Applied Research and Production Techniques

- Processing Technologies for Milk and Milk Products: Methods, Applications, and Energy Usage
- Engineering Interventions in Agricultural Processing
- Technological Interventions in the Processing of Fruits and Vegetables
- Technological Interventions in Management of Irrigated Agriculture
- Engineering Interventions in Foods and Plants
- Technological Interventions in Dairy Science: Innovative Approaches in Processing, Preservation, and Analysis of Milk Products
- Novel Dairy Processing Technologies: Techniques, Management, and Energy Conservation

ABOUT SENIOR EDITOR-IN-CHIEF

 Megh R. Goyal, PhD, PE, is a Retired Professor in Agricultural and Biomedical Engineering from the General Engineering Department in the College of Engineering at University of Puerto Rico – Mayaguez Campus; and Senior Acquisitions Editor and Senior Technical Editor-in-Chief in Agricultural and Biomedical Engineering for Apple Academic Press Inc.

He received his BSc degree in Engineering in 1971 from Punjab Agricultural University, Ludhiana, India; his MSc degree in 1977 and PhD degree in 1979 from the Ohio State University, Columbus; his Master of Divinity degree in 2001 from Puerto Rico Evangelical Seminary, Hato Rey, Puerto Rico, USA.

Since 1971, he has worked as Soil Conservation Inspector (1971); Research Assistant at Haryana Agricultural University (1972–1975) and the Ohio State University (1975–1979); Research Agricultural Engineer/Professor at Department of Agricultural Engineering of UPRM (1979–1997); and Professor in Agricultural and Biomedical Engineering at General Engineering Department of UPRM (1997–2012). He spent a one-year sabbatical leave in 2002–2003 at Biomedical Engineering Department, Florida International University, Miami, USA.

He was the first agricultural engineer to receive a professional license in agricultural engineering in 1986 from the College of Engineers and Surveyors of Puerto Rico. On September 16, 2005, he was proclaimed as "Father of Irrigation Engineering in Puerto Rico for the twentieth century" by the ASABE, Puerto Rico Section, for his pioneering work on micro irrigation, evapotranspiration, agroclimatology, and soil and water engineering. During his professional career of 45 years, he has received awards such as Scientist of the Year, Blue Ribbon Extension Award, Research Paper Award, Nolan Mitchell Young Extension Worker Award, Agricultural Engineer of the Year, Citations by Mayors of Juana Diaz and Ponce, Membership Grand Prize for ASAE Campaign, Felix Castro Rodriguez Academic

Excellence, Rashtrya Ratan Award and Bharat Excellence Award and Gold Medal, Domingo Marrero Navarro Prize, Adopted Son of Moca, Irrigation Protagonist of UPRM, Man of Drip Irrigation by Mayor of Municipalities of Mayaguez/Caguas/Ponce and Senate/Secretary of Agriculture of ELA, Puerto Rico.

He has authored more than 200 journal articles and textbooks: *Elements of Agroclimatology* (Spanish) by UNISARC, Colombia; and two bibliographies on drip irrigation. Apple Academic Press Inc. (AAP) has published his books, namely *Management of Drip/Trickle or Micro Irrigation* and *Evapotranspiration: Principles and Applications for Water Management*.

During 2014–2015, AAP has published his ten-volume set in *Research Advances in Sustainable Micro Irrigation*. During 2016–2017, AAP will be publishing book volumes on emerging technologies/issues/challenges under the book series, *Innovations and Challenges in Micro Irrigation*, and *Innovations in Agricultural and Biological Engineering*. Readers may contact him at: goyalmegh@gmail.com.

ABOUT CO-EDITOR

Dr. R. K. Sivanappan received his B.E. degree in Civil Engineering in 1953 from Madras University, India; and his M. Tech. degree in 1962 from the Indian Institute of Technology (IIT) Kharagpur, India. He was conferred an honorary doctorate degree by the Linkoping University Sweden during 1989 for his outstanding research on water management. The Tamil Nadu Agricultural University (TNAU) awarded him the Doctor of Science (Honoris causa) for his outstanding services in irrigation management in 2005.

Since 1953, he has worked as a civil engineer in the Public Works Department in Tamil Nadu State; and as Lecturer, Professor, and Dean of College of Agricultural Engineering at TNAU, Coimbatore, India. He was the Founder Director of the Water Technology Centre at TNAU, Coimbatore, by obtaining a sizeable grant from SIDA, Sweden, for his excellent services (Research and Extension Works) in irrigation water management, and soil and water conservation, especially for his contribution in the field of drip irrigation. He is known as the "Father of drip Irrigation in India", as mentioned by Sandra Postel, USA, in her well-known publication (book) *Pillar of Sand* (1999). He retired from service in 1986 after serving for a period of 33 years.

In 1968–1969 as a USAID fellow, Dr. Sivanappan went to the USA to study irrigation, drainage, soil and water conservation techniques and took advanced courses in the above subjects at the University of California, Utah State University, Oklahoma State University, and United States Soil Salinity Lab at Riverside, CA – USA.

Since his retirement from TNAU in 1986, he has been serving as a Consultant/Adviser to the World Bank, U.N.D.P. FAO, SIDA, DANIDA, JICA/JBIC, USAID, World Water International, and International Development Enterprises (IDE, USA). Within India, he has been a consultant to the Govt. of India, State Governments in India, National Consultancy companies and IPCL, AFC, NABARD, CAPART, WAPCO, INCID, and many more. He has worked as a member in the State Planning Commission, Tamil Nadu (2005–2006), incharge of water resources, irrigation, land use, etc.

He has received more than 10 awards from different organizations both in India and abroad, including from Indian Society of Agricultural. Engineers, New Delhi., Soil conservation society of India, New Delhi, The Institution of Engineers (India) Calcutta, and other professional societies.

He has authored about 30 books in irrigation, water management, dryland farming, sprinkler irrigation, drip irrigation, farming in wind farm lands, soil and water conservation, waste land and fallow land development, water harvesting, agro forestry in dryland, and Tamil Nadu Water Vision and Interlinking of Indian Rivers, etc. He has published more than 1000 journal articles in national and international journals; and about 50 manuals and reports on water and water-related topics. He has traveled to more than 30 countries for consultancy work and for attending international conferences, seminars/workshops.

Since 2012, he has been as a book contributor to Apple Academic Press Inc., and is a frequent contributor to the book volumes under drip irrigation series, and AAP's Agricultural and Biological Engineering book series. Today at the age of 87 years, he is continuing his services to help the farmers and the education community. Readers may contact him at: sivanappanrk@hotmail.com.

EDITORIAL

Apple Academic Press Inc. (AAP) will be publishing various book volumes on the focus areas under book series titled *Innovations in Agricultural and Biological Engineering*. Over a span of 8 to 10 years, Apple Academic Press Inc., will publish subsequent volumes in the specialty areas defined by the *American Society of Agricultural and Biological Engineers* (http://asabe.org). We seek book proposals from the readers in areas of their expertise.

The mission of this series is to provide knowledge and techniques for agricultural and biological engineers (ABEs). The series aims to offer high-quality reference and academic content in Agricultural and biological engineering that is accessible to academicians, researchers, scientists, university faculty, and university-level students and professionals around the world.

The following material has been edited/modified and reproduced below (*Goyal, Megh R., 2006. Agricultural and biomedical engineering: Scope and opportunities. Paper Edu_47 at the Fourth LACCEI International Latin American and Caribbean Conference for Engineering and Technology (LACCEI' 2006): Breaking Frontiers and Barriers in Engineering: Education and Research by LACCEI University of Puerto Rico – Mayaguez Campus, Mayaguez, Puerto Rico, June 21–23*):

WHAT IS AGRICULTURAL AND BIOLOGICAL ENGINEERING (ABE)?

"Agricultural Engineering (AE) involves application of engineering to production, processing, preservation and handling of food, fiber, and shelter. It also includes transfer of technology for the development and welfare of rural communities," according to http://isae.in. *"ABE is the discipline of engineering that applies engineering principles and the fundamental concepts of biology to agricultural and biological systems and tools, for*

the safe, efficient and environmentally sensitive production, processing, and management of agricultural, biological, food, and natural resources systems," according to http://asabe.org. *"AE is the branch of engineering involved with the design of farm machinery, with soil management, land development, and mechanization and automation of livestock farming, and with the efficient planting, harvesting, storage, and processing of farm commodities,"* definition by: http://dictionary.reference.com/browse/agricultural+engineering.

"AE incorporates many science disciplines and technology practices to the efficient production and processing of food, feed, fiber and fuels. It involves disciplines like mechanical engineering (agricultural machinery and automated machine systems), soil science (crop nutrient and fertilization, etc.), environmental sciences (drainage and irrigation), plant biology (seeding and plant growth management), animal science (farm animals and housing) etc.," by: http://www.ABE.ncsu.edu/academic/agricultural-engineering.php.

"According to https://en.wikipedia.org/wiki/Biological_engineering: *"Biological engineering (BE) is a science-based discipline that applies concepts and methods of biology to solve real-world problems related to the life sciences or the application thereof. In this context, while traditional engineering applies physical and mathematical sciences to analyze, design and manufacture inanimate tools, structures and processes, biological engineering uses biology to study and advance applications of living systems."*

SPECIALTY AREAS OF ABE

Agricultural and Biological Engineers (ABEs) ensure that the world has the necessities of life including safe and plentiful food, clean air and water, renewable fuel and energy, safe working conditions, and a healthy environment by employing knowledge and expertise of sciences, both pure and applied, and engineering principles. Biological engineering applies engineering practices to problems and opportunities presented by living things and the natural environment in agriculture. BA engineers understand the interrelationships between technology and living systems, have available a wide variety of employment options." *ABE embraces a vari-*

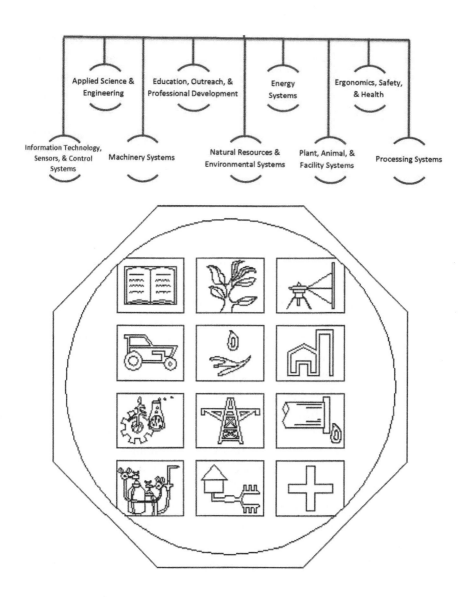

ety of following specialty areas," http://asabe.org. As new technology and information emerge, specialty areas are created, and many overlap with one or more other areas.

1. **Aquacultural Engineering**: ABEs help design farm systems for raising fish and shellfish, as well as ornamental and bait fish. They

xxxvi   Editorial

specialize in water quality, biotechnology, machinery, natural resources, feeding and ventilation systems, and sanitation. They seek ways to reduce pollution from aquacultural discharges, to reduce excess water use, and to improve farm systems. They also work with aquatic animal harvesting, sorting, and processing.

2. **Biological Engineering** applies engineering practices to problems and opportunities presented by living things and the natural environment.

3. **Energy:** ABEs identify and develop viable energy sources – biomass, methane, and vegetable oil, to name a few – and to make these and other systems cleaner and more efficient. These specialists also develop energy conservation strategies to reduce costs and protect the environment, and they design traditional and alternative energy systems to meet the needs of agricultural operations.

4. **Farm Machinery and Power Engineering**: ABEs in this specialty focus on designing advanced equipment, making it more efficient and less demanding of our natural resources. They develop equipment for food processing, highly precise crop spraying, agricultural commodity and waste transport, and turf and landscape maintenance, as well as equipment for such specialized tasks as removing seaweed from beaches. This is in addition to the tractors, tillage equipment, irrigation equipment, and harvest equipment that have done so much to reduce the drudgery of farming.

5. **Food and Process Engineering:** Food and process engineers combine design expertise with manufacturing methods to develop economical and responsible processing solutions for industry. Also food and process engineers look for ways to reduce waste by devising alternatives for treatment, disposal and utilization.

6. **Forest Engineering**: ABEs apply engineering to solve natural resource and environment problems in forest production systems and related manufacturing industries. Engineering skills and expertise are needed to address problems related to equipment design and manufacturing, forest access systems design and construction; machine-soil interaction and erosion control; forest operations analysis and improvement; decision modeling; and wood product design and manufacturing.

7. **Information and Electrical Technologies Engineering** is one of the most versatile areas of the ABE specialty areas, because it is applied to virtually all the others, from machinery design to soil testing to food quality and safety control. Geographic information systems, global positioning systems, machine instrumentation and controls, electromagnetics, bioinformatics, biorobotics, machine vision, sensors, spectroscopy: These are some of the exciting information and electrical technologies being used today and being developed for the future.

8. **Natural Resources:** ABEs with environmental expertise work to better understand the complex mechanics of these resources, so that they can be used efficiently and without degradation. ABEs determine crop water requirements and design irrigation systems. They are experts in agricultural hydrology principles, such as controlling drainage, and they implement ways to control soil erosion and study the environmental effects of sediment on stream quality. Natural resources engineers design, build, operate and maintain water control structures for reservoirs, floodways and channels. They also work on water treatment systems, wetlands protection, and other water issues.

9. **Nursery and Greenhouse Engineering**: In many ways, nursery and greenhouse operations are microcosms of large-scale production agriculture, with many similar needs – irrigation, mechanization, disease and pest control, and nutrient application. However, other engineering needs also present themselves in nursery and greenhouse operations: equipment for transplantation; control systems for temperature, humidity, and ventilation; and plant biology issues, such as hydroponics, tissue culture, and seedling propagation methods. And sometimes the challenges are extraterrestrial: ABEs at NASA are designing greenhouse systems to support a manned expedition to Mars!

10. **Safety and Health:** ABEs analyze health and injury data, the use and possible misuse of machines, and equipment compliance with standards and regulation. They constantly look for ways in which the safety of equipment, materials and agricultural practices can

be improved and for ways in which safety and health issues can be communicated to the public.

11. **Structures and Environment:** ABEs with expertise in structures and environment design animal housing, storage structures, and greenhouses, with ventilation systems, temperature and humidity controls, and structural strength appropriate for their climate and purpose. They also devise better practices and systems for storing, recovering, reusing, and transporting waste products.

CAREER IN AGRICULTURAL AND BIOLOGICAL ENGINEERING

One will find that university ABE programs have many names, such as biological systems engineering, bioresource engineering, environmental engineering, forest engineering, or food and process engineering. Whatever the title, the typical curriculum begins with courses in writing, social sciences, and economics, along with mathematics (calculus and statistics), chemistry, physics, and biology. The student gains a fundamental knowledge of the life sciences and how biological systems interact with their environment. One also takes engineering courses, such as thermodynamics, mechanics, instrumentation and controls, electronics and electrical circuits, and engineering design. Then the student adds courses related to his particular interests, perhaps including mechanization, soil and water resource management, food and process engineering, industrial microbiology, biological engineering or pest management. As seniors, engineering students team up to design, build, and test new processes or products.

For more information on this series, readers may contact:

Ashish Kumar, Publisher and President	Megh R. Goyal, PhD, PE
Sandy Sickels, Vice President	Book Series Senior Editor-in-
Apple Academic Press, Inc.,	Chief
Fax: 866-222-9549	Innovations in Agricultural and
E-mail: ashish@appleacademicpress.com	Biological Engineering
http://www.appleacademicpress.com/	E-mail: goyalmegh@gmail.com
publishwithus.php	

PART I

SOIL AND WATER CONSERVATION AND MANAGEMENT

CHAPTER 1

AGRICULTURAL WATER MANAGEMENT: CHALLENGES IN RESEARCH AND DEVELOPMENT

R. K. SIVANAPPAN

CONTENTS

1.1 INTRODUCTION

1.1.1 IMPORTANCE OF WATER

The steeply ascending curve which depicts the dramatic rise in world's population is now very familiar to agricultural scientists and others concerned in meeting man's future needs. It is a fact that as man's standard of living improves, the per capita demand for water also increases sharply. Thus, if man continues to multiply as projected and to use water as forecast based on today's trends, the curve forecasting for water requirements much ascend even more steeply than that for population.

The total supply of water on this planet earth is essentially fixed. Thus, man must find ways to increase crop production and also meet his other requirements for water by more efficient use of world's limited water supply. The earth's reservoir of water is largely the vast oceans. One fourth of the sun's total energy is used to evaporate approximately three feet of pure water from the seas. A portion of this water vapor moves over land masses and falls as rain or snow. The term hydrological cycle is used to describe the movement of water in its several states through various processes in the passage from the oceans to the atmosphere and its return to the seas. The portion of the cycle important for water utilization by man is that between the receipt of precipitation on the earth and the return of water directly to the air or by runoff to the sea. The practical limits of man's ability to influence this phase of the hydrologic cycle are determined by physical and technologic restrictions and by economic, legal and social problems.

Further, the importance and scarcity of water can be seen from the fact that the availability of water is about 6000–7000 cum/person/year globally, compared to 2000 cum/person/year in India. In some States like Tamil Nadu the availability is only about 650 cum/person/year. But the minimum requirement of water for a reasonable living for an individual basis is about 1000 cum/year, "*A country whose renewable fresh water availability exceeds 1700 m³/person/year will suffer only occasional or local water problems. Below this threshold, countries begin to experience periodic or regular water stress. When fresh water availability falls below 1000 m³/person/year, countries experience water scarcity in which the lack of water begins to hamper economic development and human health of well-being. When renewable fresh water supplies fall below 500 m³/person/year, countries experience absolute scarcity. The 1000 m³ bench mark has been accepted as a general indicator of water scarcity,"* according to report by World Bank.

1.1.2 WATER USE

Irrigation presently takes large percentage of the water supplies of arid regions. More than 90% of water is diverted for irrigation in most developing countries. But when the demand for water is increasing, the allocation of water for agriculture is being reduced.

It is mentioned that though the area of irrigation is going to be doubled in USA, the additional requirement was expected to increase only by 2% since the research findings in water management is increasing and the knowledge will be put into operation by the extension personnel. Similarly, the allocation of water for agriculture in Israel was about 85% in 1965, the same was reduced to 75% in 1975 and about 65% in 1985 or so. But the area of irrigation is increasing due to the advancement made in water management practices. The scientists have indicated that the allocation of water for agriculture is more than 90% in India and this will come down to about 75% in another 15 to 20 years. However, to feed the increased population, it is necessary to bring more area under irrigation in the next 20 years. This needs more information on the economy in the use of water in agriculture.

1.1.3 DEMAND FOR WATER

India is blessed with good rainfall and water resources. The average rainfall is about 1150 mm compared to world average of about 840 mm. The sunshine is also very favorable to take crops all-round the year in most of the States. Though the irrigated area in 1950–51 was only about 21 M.M.Ha and this has been increased to 65 M.Ha in 1984, and to more than 100 M.Ha in 2012. The food production has also increased from 50 MT to 250 MT during this period. To feed the expected population of 1650 million in the year 2050, the food production target is fixed at 450 million Tons. To achieve this goal, it is necessary to bring more area under irrigation from the available water resources.

It is estimated that the entire harnessable surface water and rechargeable ground water could be fully utilized by construction of dams and channels and open and tube wells, respectively by 2020. The area that can be brought under irrigation will be about 115 M.Ha which is about 52% of the total cultivable area at that time. What is to be done afterwards? The demand for water will be increasing many folds not only for agriculture, but also for industries and for municipal and drinking purposes. On the one hand the allocation of water for agriculture will have to be reduced and on the other side, more area should be brought under irrigation to feed the growing population. This needs intensive and extensive research in better utilization of water.

1.1.4   INCREASING WATER SUPPLY

The increasing scarcity of water and its implication for the economy of the country have become one of the most concerns of the Governments and it may be expected that this will increase in the years to come. The lessons derived from the study of the present situations show that only by investing a major effort in water saving, it will be possible to maintain a reasonable growth rate of that part of the national income dependent on irrigated agriculture in the areas where water is a limiting factor. Hence, there is immediate need to find out ways and means to increase the water supply. Since the rainfall and availability of water resource is constant, the alternatives have to be found out for the increased demand. The following are some of the measures which will help in increasing the irrigated area in the coming years:
- Desalination.
- Salt water utilization.
- Reuse of waste water (Effluent and sewage water).
- Weather modification.
- Improved water management practices.
- Water harvesting and conservation through watershed development.

1.1.4.1   Desalination

More than 97% of the water on the planet earth is saline and cannot be used for agriculture. Even the fraction of water available, it is not evenly distributed and hence scarcity is noticed in many places. If the available salt water could be reclaimed by desalination, within reasonable cost, it may be possible to obtain the required quantity of water. Though the cost of desalination per unit quantity of water is too much at the present available technology, there is a possibility to reduce the cost by applying the latest technology through more research in that direction. The problem is that the reclaimed water is to be pumped to reach the interior areas for irrigation pumps, which will add further cost. Instead, the desalinated water could be used for drinking and municipal purposes in the coastal population and for the industries in these regions. The water used for these purposes can be diverted to agriculture.

The survey conducted has shown that there were more than one thousand land based plants and many more are under construction. A major problem associated with desalination is how to dispose of the salt removed from the water. Another problem is that it requires substantial amount of energy. Because of increasing concern about energy supply, energy costs and allocation of available energy, decisions pertaining to desalination on a methodology of increasing available water supplies must be reached only after careful study and research.

1.1.4.2   Salt Water Utilization

In many areas, water is available in the underground, but it could not be used for growing crops since it is saline and have more salt contents. Scientists are doing research to find out the management practices to utilize the salt water for raising crops. Further, crops are being identified which can be grown in different situations. Drip irrigation is one of the methods by which the salt water can be used for growing crops to save water.

1.1.4.3   Reuse of Water

Large quantity of water is diverted for municipal needs and for industrial purposes. This quantity may go up to 30% of the total available water in the next 20–25 years. The entire water is not consumed by human beings or by the industry, more than 90% is left out as waste water. Hence, the used water or waste water can be collected and profitably used after some treatment. Even today, some municipalities are collecting their sewage water and after minor treatment, utilize the same for growing fodder crops. Similarly, the industrial waste water is being treated and used for agriculture in some industries. But this should be properly regulated and used to the maximum extent since substantial area could be brought under irrigation by utilizing this water. In advanced countries like USA, Israel, Australia, the sewage water is reclaimed and used even for drinking purposes after recharging it into ground through water spreading basins. The reclaimed water is also used for all types of fruit crops and cotton. Israel now reuse more than 70% of its domestic waste water for crop production. Treated waste water accounts for 30% of the nations agricultural water supply and this quantity is expected to reach more than 80% in another 10 years.

1.1.4.4 Weather Modification

The seeding or artificial rain making has come into practice even in developing countries. Studies conducted in USA, Israel, Australia have indicated that by appropriate techniques, it is possible to increase the rain by 15–20%. The Indian Institute of Meteorology has succeeded in increasing rainfall by 10 to 20% in the demonstration area. However, it is cautioned that the total rainfall over the earth may be the same, but the increased rain in a particular area will enhance the water need of that area and for increasing the yield of crops in that region since the needed moisture is given by enhancing the rain. What lies ahead for improved application of cloud seeding is a future that depends on the past on the extent that society is willing to invest further in scientific learning and in part on the effort is made to create social institutions capable of reconciling the frequently conflicting desires and needs of society.

1.1.4.5 Improved Water Management Practices

The yet another way of increasing the irrigated area from the available water is by adopting better water management practices. Irrigation is as old as civilization and the same age old method of irrigation is being followed even today. Scientists have developed techniques/methods by which large quantity of water could be saved and yield could be increased. This is possible by altering the time and special distribution of water through well planned, adequately financed, properly constructed and efficiently managed irrigation and other water projects. Irrigation projects are complex development enterprises which require the systematic and efficient harvest of water, its conveyance and application in controlled quantities to suitable areas of land in an integrated manner of a large number of complementary inputs essential to increase or even to allow crop production.

1.1.4.6 Water Harvesting Conservation and Management

Water is essential for life and used in many different ways. It is also a part of the large eco system on which the production of biodiversity depends. Fresh water scarcity is not limited to the arid regions only; in areas with good supply, access to safe water is becoming a critical problem. Lack of water is caused by low water storage capacity, low infiltration, large inter annual and annual fluctuations of precipitation due to monsoon rains and high evaporation demand.

Rain water harvesting is as old as civilization and has been practiced continuously in different ways for different purposes in the world. Nonetheless, it has not been systematically in all places. A need has come to harvest rain water including roof water, to solve water problems everywhere, not only in arid but also in humid regions.

1.2 AGRICULTURAL WATER MANAGEMENT

Agricultural water management may be defined as a skill of coordination of water resource, tapping, receipt, storage, conveyance, diversion, delivery, distribution and application constituent with the soil capability and the crop requirements for maximizing irrigation efficiency and economic returns. This definition implies resource inventory, system analysis, decision making and project evaluation. Hence, the scope of water management depends on the inter-relationships among variety of factors including irrigation practices, land use and cropping pattern individual and collection system of irrigation and overall economics. The definition of water management incorporates the critical role of risk factor, a new dimension which is well defined in other business disciplines, but not in agriculture. This dimension has a direct impact on nation's economy. The risk factor associated with farm irrigation management is largely due to the erratic behavior of monsoon, the imbalance created in the supply – demand structure of agricultural commodities and the consequential variations in the marketing price, Government price fixation policy and the overall stress the farmer being subjected to by factors other than agriculture.

1.2.1 COMPLEX NATURE OF WATER MANAGEMENT

The various activities in water management practices will explain the complex and complicated nature of works involved in achieving the objectives. Further, the attainment of saving in water by implementing various water management practices in the field is extremely a complex one since the storing and distribution of water is looked after by the Public Works Department, the on-farm development including leveling, providing channels are done by the Department of Agricultural Engineering and the Departments of Agriculture, Co-operation and Revenue are involved in one way or other in the field for raising crop and using the water. There are numerous farmers involved in these activities and they should also cooperate in utilizing water based on the availability and advice given by the extension agency. The saving is dependent as the individual's awareness of the need to control and economize in water use.

The cost of water is not based on the quantity of water used in tank/canal irrigated areas and the farmers can use any amount whether it is needed or not. In many cases excess water is used and it is neither controlled nor regulated to get the required quantity of water to the crops. The individual farmer is not benefitted by saving water though it may be beneficial for the State. In order to use improved water management practices and to save water, a national feeling should come to the individual farmer. Hence, it is a very complex in nature and difficult to solve the problem.

1.2.2 NEED FOR WATER MANAGEMENT IN INDIA

India is a vast country with varied agro-climatic regions with different soil, rainfall and other ecological aspects. The diversification of crops in various regions depending on soil and different elevations need different water management practices. The topography of the country especially in river basin is such that, it is difficult to store all the rain water when it falls within three to four months. Within the limits of available technology, it appears possible to utilize only about 50% of the average annual surface flow and 75% of the groundwater recharge economically. This quantity has to be enhanced for the coming years in order to cope up with the demand of water. The picture in Tamil Nadu is different where already the available surface and ground water resources have been harnessed. To bring more area under irrigation, the only possible way is to economize the use of water for all crops especially for paddy which takes more than 45% of the agricultural water used for irrigating 30% of the total irrigation area in the country. But in Tamil Nadu and Andhra Pradesh, more than 80% of water is given for rice crop. There are numerous ways and means by which water conservation and management could be achieved. In Kerala, though the rainfall and availability of water are plenty in monsoon sea-son (July–October), but the water scarcity is experienced in other months. This is further noticed in hills where plantation crops are grown using the rain as main source of irrigation, but the plants suffer for want of water during December–May and if the water is properly managed by giving it throughout the year, the income of the farmers can be increased substanti-ality. Therefore, the need for water management is there everywhere and every place.

An efficient water management system aims to reduce water con-sumption per unit of product to minimal. Reduction in consumption could be achieved either by physical means and engineering construction and distribution or by a system of economic disincentives for wasteful uses and incentive, for economic uses or by mixer of these methods in various proportions. With water scarcities facing the country, it is essen-tial that agriculturists should become keenly aware of the real value of water and its economic use especially the farmers using the canal and tank irrigation.

1.2.3 ON FARM DEVELOPMENT AND MANAGEMENT OF WATER

The term 'On Farm Development' means the development of individual field to use the canal/tank water very efficiently. The works involved are farming irrigation and drainage channels providing measuring devices, control and regulation structures, land leveling and shaping, etc. The main, branch channel and distributors should be designed based on the peak water use and the sluices should be constructed to allow the required quantity of water whenever needed.

The term farm water management like irrigation management and similar expressions is confusing since it means different things to different people. Since bringing more water to the farms does not yield the expected increase in food production, there are many who believe it is the farmer who is to be blamed. The farmer is considered backward and not to know how to make efficient and economic use of his water. The term farm water management is needed at the farm level. Whatever the causes of the problems, it is at this level that food is produced and to raise production, there is a great need to make better use of available water. Farm level, in this sense, refers to individual fields or holdings as well as to elements of the water distribution system (e.g., the water course) whose operation and maintenance is primarily the responsibility of the farmers.

However, it has become clear that farmers are unlikely to irrigate more efficiently as long as the supply of water to their farms at the right time and right rate is not reliable. Achieving full reliability of supplies and integration of main system operations with those at the farm level is a pre-requisite for efficient water use by the farmers. Therefore, farm water management as a whole concept should refer not only to the farm level, but should also cover the operation of the main irrigation system. Furthermore, management implies the involvement of people, farmers, system operators, village administrators and Government officials. The attitudes and interests of all involved, as well as their inter-relationships, interactions and organization are of vital importance to making efficient use of structural improvements in irrigation systems. It follows that research and development work in the social, economic and institutional areas, is an essential part of farm water management.

1.3   RESEARCH IN WATER MANAGEMENT

1.3.1   RESEARCH: AN INTER-DISCIPLINARY EFFORT

Water management research is an inter-disciplinary in nature involving Irrigation Engineers, Agricultural Engineers, Agronomists, Soil Scientists, Agricultural Economists (Social Scientists) and Plant Physiologists. Each one has to play an important role in finding solutions in management of water in the Project Area/Farm. The Irrigation Engineer has to study the system approach to tackle the water management problem for the whole project. The Agricultural Engineer should find out the optimum level of irrigation/drainage channel, providing necessary structures on the farm, proper land leveling and shaping. The Agronomists and Soil Scientists are to evolve suitable crop/cropping pattern, based on the availability of water, crop season, and soils. The Economist is expected to work out the economics of the scheme and its viability. The Crop Physiologist is responsible to find out answers what happen if there is water stress and how to overcome this problem. But if the scientist in each discipline is tackling the problems in an isolated manner, the impact of water management cannot be achieved especially in a big project. Therefore, all the scientists should work together to bring out the overall water management practices covering all aspects as mentioned earlier. It is very difficult to find out solutions on water management as done in the case of fertilization or pest and disease protection. Therefore, a concerted effort by a group of scientists should be vigorously attempted to, in research on water management.

1.3.2   WATER MANAGEMENT RESEARCH IN ADVANCED COUNTRIES

The research on water management has been more than 75 years old in advanced countries like USA Though water was not a problem in many countries, advanced technology in water use is being practiced for nearly 50 years. The aim was to reduce our requirement for irrigation. They developed the sprinkler irrigation as early in 1900 and much progress was made after the Second World War after the introduction of aluminum pipes and electric motors. The advanced method of irrigation namely drip system is in practice nearly 50 years. The scientists calculated the water used for various purposes and projected the need for the next 20–25 years. Further, they worked out the details of how the demand is going to be met with. Side by side, extension side on water management was established in these countries and the improved methods were popularized in large scale in the fields. The area of irrigation by sprinkler and drip was increased and it is now about 20% in USA, 90% in Israel and in large areas in Mediterranean countries in Europe. After the introduction of drip Irrigation, the same is popular among farmers in going for orchard and vegetable crops. Nearly six million farms are being irrigated by drip method in the world. They also introduced drainage (surface and sub surface) wherever it is necessary to avoid salt accumulation and turning into problem soils. The trend is continuing and more importance is given to research on water management including drainage.

1.3.3   COORDINATED RESEARCH IN WATER MANAGEMENT IN INDIA

The coordinated research in water management was initiated by the Indian Council of Agricultural Research (ICAR) during 1968 or so with limited river valley projects. There are about 35 research schemes scattered in the entire country. Apart from that Agricultural Universities, Engineering Institutions are having their own research programs on water management. Many institutions have established to take up studies only on water management namely, Water Technology Center, New Delhi, Center for Water Resources Development and Management, Calicut, and Water Technology Center, Coimbatore and Bhubaneswar. Five Water Management Training Institutions with the help of USAID to train personnel on water management in the Department of Agriculture. Irrigation and Agricultural Engineering were sanctioned and these institutions will take up action research or live research in a large scale which will be acting as Laboratory to the trainees in water management.

1.3.4   METHODOLOGY IN RESEARCH

In the earlier days, water management studies were done on the water requirements of crops. The methodology/techniques followed are irrigating the crop after allowing various moisture levels of depletion in the soil depending upon the crop. Based on the studies, when to irrigate and how much water should be given were determined. Recently, the water requirements studies are taken based on the IW/CPE ration again depending upon the crop varying from 0.4 to 1.0 giving the fixed quantity of water ranging 4 or 5 or 6 cm. This will help in finding out when to irrigate for various crops and how much water. Enough basic research is available about the water requirements of various crops. The problem is to use the findings in an integrated manner on the requirements of water for land preparation, deciding the size of the basins or length of furrows for improving irrigation efficiency. Reclamation of problem soils, use of saline water, water production functions are some of the works undertaken by the various research workers.

1.4   RESEARCH AREAS IN WATER MANAGEMENT

The area of research in Agricultural Water Management is vast. Research is being carried out in conserving and utilizing the available water in the various situations. The areas can be classified as follows:
    a.   Increasing precipitation and utilization of runoff from watershed
    b.   Water storage
    c.   Water conveyance
    d.   Irrigation system and land preparation
    e.   Water quality studies and use the saline water
    f.   Problem soils and reclamation
    g.   Irrigation management
    h.   Drainage need and methods

1.4.1   INCREASING PRECIPITATION AND UTILIZATION OF RUNOFF FROM WATERSHED

Great efforts have been made by the scientific community and by commercial operators to develop a better understanding of weather mechanisms and techniques for altering conditions as to create precipitation on demand. These efforts have led to much excitement and conflicting information's. Numerous reports have concluded that it is possible to increase the rainfall by about 15–20% and much more effort is needed in this direction.

The elaborate water harvesting systems pain stakingly built by the Nabateans and others in the Negev Desert region of the Middle East before the time of Christ, provide the basis for renewed research efforts seeking improvement in man's ability to concentrate rainfall of arid regions sufficiently to obtain useful plant growth. Although the basic idea is ancient, this new research, which may well be pioneering new technology, is underway in Israel, Australia, Northern Mexico and other locations.

Precipitation falling on the watershed is partially retained by the vegetative cover and lost by evaporation to the atmosphere. This interception loss varies widely depending on climate and season, plant species and density of cover and other factors. In terms of percentage of annual precipitation, interception losses may range from less than 5% to over 50%. Also the vegetation growing on the watershed uses water from the soil and rock mantel below. In the arid areas, the watershed vegetation may remove most of the water from the soil and also form deep formations. Water deficits created by the vegetative transpiration must be replaced by subsequent precipitation. There have been intensive studies of possibilities for managing the vegetative cover of watersheds so as to control runoff and increase the water yields. This often involves replacement of deeprooted plants with shallow rooted types to help reduce depletion of water from the soil and rock mantel. Much attention has been given to riparian vegetation and pyreatrophytes. It appears that removal of these plants can save substantial quantities of water which may be of great importance in water deficit areas.

1.4.2 WATER STORAGE

Where surface or ground waters are available for development and use for irrigated agriculture, studies are necessary to determine the most feasible method of storage and utilization. Where the water supplies are the limiting factor, economics of dam construction may be secondary to efficiency of water storage.

In arid areas of the world, evaporation from water surfaces may be as great as 3 meter in depth. Where evaporation is great, primary consideration must be given to reservoir sites which produce an area of minimum exposure to evaporation. Application of cetyl alcohol compounds to form a monomolecular layer on the water surface retards evaporation. Laboratory tests show that as much as 65% of evaporation can be eliminated under controlled conditions by use of this material. However, success in using this approach depends entirely upon the conditions existing at the reservoir. Further research will be needed to overcome the problems of wind and wave action. Attention also needs to be given to the effects of evaporation control measures on aquatic life.

Control of aquatic vegetation is also an important factor in operations in order to decrease the losses by non-economic transpiration. Vegetation control by manipulation of reservoir levels, chemical treatment, burning mechanical removal are all utilized in this control. Seepage from small reservoirs farm ponds can be controlled by lining the reservoir with concrete, plastics, clay, salt and other chemicals.

Ground water basins should be operated to maintain water levels in order to hold pyreatophytic vegetation to a minimum, and to maintain water levels at depth to overcome drainage and salinity problems. Where feasible, conjunctive use of surface and ground water reservoirs can be programmed to maximize available supplies and minimize loss. Depleted ground water supplies can be replenished by artificial means. Artificial recharge can be accomplished by diverting water from natural streams, channels and spreading it over adjacent permeable soils, by releasing water stored in surface reservoirs to natural channels during periods of low flow, or by injecting water into wells, shafts or pits. These operations all encounter difficulties which will require more pioneering research for their solution.

1.4.3 WATER CONVEYANCE

With conveyance losses up to 50% causing loss of valuable water and the water logging and salinization of nearly valuable crop land, serious attention must be given to finding economical methods to reduce all forms of losses. Research findings have indicated that even in garden land of small farm condition, the conveyance losses are estimated to about 20%. Research is being carried out towards development of more practical and means of materials for conveying water to fields on the farm. Such equipment must be simple, reliable and economical if it is to be accepted and used to achieve more efficient water management. The farmers are going for underground pressure/non-pressure pipes or prefabricated channels with cement mortar for taking water from one place to other place depending upon the situations. PVC pipes and accessories are available which can be used in conveying the water. Unless the water is conveyed through pipes/other structures without any seepage loss, it will be difficult to control, regulate and provide the required quantity of water. Number of control and regulating structures are available and the same can be used for better water management. It will also help to achieve efficient irrigation, optimize crop production and minimize hazards from water logging and resultant salinization of soil. Research work continues on better economical channel design, improved material and equipment and more efficient management of the system.

1.4.4   IRRIGATION SYSTEMS AND LAND PREPARATION

In many modern irrigation projects, farmers continue to irrigate in accordance with age-old traditions using equipment and practices little influenced by modern science and technology. Irrigation was practiced extensively by the earliest civilizations known. Much of this irrigated agriculture ultimately failed because of technical problems created by incomplete planning and by misuse of water and soils. These problems included the application of insufficient irrigation water under conditions where salts could accumulate or the application of excessive depths of irrigation water which leached out soil nutrients, caused ground water tables to rise, and ultimately caused the accumulation of toxic salts. Unfortunately, the same problems continue to cause the failure of some recently established irrigation projects, reduce crop yields far below potential productivity and create economic and social unrest which so often accompanies slowly developing and unproductive irrigation projects, It is tragic that even today farmers in some areas are unprepared to make efficient use of water which has been supplied to them at great expense. In some areas, food production lost through deterioration of irrigated lands through mismanagement is estimated to offset production from additional lands being brought into production under costly new irrigation schemes.

Throughout the irrigated areas of the world, often twice as much water is delivered to the farm than is required by the crop to produce maximum yields. The excess water is usually lost by deep percolation below the root zone or in surface runoff. Although seepage and deep percolation losses are important to the immediate water users, this water may later be recoverable from groundwater reservoirs. Deep percolation beneath the plant root zone may be essential to the maintenance of a favorable salt balance in the soil. This is especially true where irrigation water is high in salts. Even when irrigation water is low in salts, difficulties may develop over a period of years unless sufficient water is percolated through the root zone to remove the soluble salts.

In order to distribute the water efficiently by surface irrigation, it is necessary that the land be graded to the proper slopes both in the direction of flow and across the slopes. Research studies have indicated that by leveling alone, about 30% of water could be saved in addition to saving of labor. More research attention should be given to the development of special land preparation techniques to improve water use and increase crop production in given situations.

Further, the preparation of land including the size of basins and length of furrows are not properly considered. The farmers use arbitrary size of basin without considering the soil, depth, and discharge of the source. Sometimes big size basins decrease the irrigation efficiency and by farming small basins (checks) the bunds occupy more area and thereby production may be reduced. The land preparation techniques will differ from area to area, crop to crop and for the different irrigation sources namely canal, tank and well. All these factors are to be considered and suitable systems and land preparation should be followed.

1.4.5 WATER QUALITY

As man makes more intensive use of water resources, the presence of dissolved and suspended materials in water becomes strikingly more important. Fortunately, man has now come to recognize that water management programs must be given attention to both the quantity and the quality of water.

Appropriate to mention here is pioneering research directed to understand water movement through porous materials, interactions between solutes in water and soil or other materials and the effects of solutes on the soil and on plant growth. Such research opens the way to more efficient operations for leaching excess salts from soils, predicting the suitability of given waters for irrigation or other use, adjusting quantities of irrigation water applied so as to achieve favorable salt balance and predicting the composition of water percolation in to ground water basins or returning to rivers and lakes. Here again modern techniques of model analysis and the use of computers are greatly expanding man's ability to predict effects of using waters and also resultant changes in their composition.

An interesting proposal to permit use of salty waters in irrigation agriculture has been made by Miller. Using plastic sheeting supported over a furrow carrying salty water between two rows of germinating plants, he creates a hydrologic mini-cycle. The plastic forms of solar still which evaporates pure water from the salty irrigation water which then condenses on the plastic and runs down to wet the soil around the young plants. Later when the young plants become established, they can tolerate direct applications of the saline irrigation water. Evaporation losses are also reduced while the plastic canopy is in use. Preliminary reports indicate the cotton yield was raised from 2 to 2.50 bales per acre while water use decreased from 41 to 251/2 inches.

Scientists have identified the type of crops which can be grown from the saline water depending upon the soil and the area. Further studies are underway to develop cropping pattern which will permit the most efficient use of water supplied at essentially a constant rate throughout the year.

1.4.6   CHALLENGES IN RECLAMATION OF SOILS

Many areas have become saline and alkaline after irrigation was introduced. Since the farmers use water abundantly when the water is released in the canal, the water table has been raised and in some cases the water table is above the surface especially during rainy season. Large areas have become uncultivable and hence attention was paid to find solutions. Since fertile lands are turned to be unproductive ones, soil and water are becoming saline. Research works have been taken up to identify the problem area whether it is due to salinity or alkalinity and also is it because of water or by man's fault. In many irrigation projects, there are no channels to remove the excess water. The fact that the total length of irrigation and drainage channel is less than 10 m/ha compared to 60 and 70 m/ha in well developed and managed irrigation system show that lot of exercise should be done in our irrigation projects to prevent this hazard. At the same time action has to be taken up to reclaim the affected one. Technology is available now, but it may be little costly. But this is worthwhile since these lands can produce more once if it is turned to be good one. Management of land and water is necessary in order to maintain the productivity. More research is required in this direction.

1.4.7   IRRIGATION MANAGEMENT

A recent report of the Food and Agriculture Organization of the United Nations suggests that improved water management (on farm fields, including irrigation and drainage practices) can probably do more towards increasing food supplies and agricultural income in the irrigated areas of the world than any other agricultural practice.

Irrigation needs and practices necessarily vary widely. This complicates the planning of new irrigation projects and also the operation of existing irrigation systems and irrigated farms. One important aspect of science is the ability to predict what can be expected in given situations. Mankind in general and particularly the less-developed countries can ill-afford the time-wasting resource-depleting and disappointing process of trial and error approaches and inadequately planned local trials. Research is leading to the development of basic principles for irrigation water management which can serve as valuable guides in predicting the suitability of proposed irrigation waters estimating crop irrigation requirements, predicting the effects of specific irrigation practices and related farm operations on given crops under prevailing site conditions, diagnosing drainage needs and estimating leaching requirements to maintain favorable salt balances. Considerable information is also becoming available on the tremendously important inter-relations between irrigation, fertilization, tillage and other crop production practices.

Opportunities to improve irrigation management rest upon better understanding of water in the soil-plant-water system. Many scientists have contributed to our present understanding of saturated and unsaturated flow of water in soils, its retention and movement to plant roots; to knowledge about absorption of water, its loss from plant leaves and its energy status and effects on metabolic processes in plants, and to our understanding of factors affecting light absorption and transpiration by leaves.

Since water, especially in the arid regions is a limiting and usually a costly resource, it is generally desirable to plan irrigation programs for 'efficiency' in terms of maximizing crop yield per unit of water applied. In some cases, it may be preferable to maximize crop yield per unit of irrigated land or per unit of initial investment in land preparation or irrigation distribution system. Thus, the most desirable irrigation practice will vary with the situation and depend upon proper integration of all factors involved. Efforts are now being made to formulate some general irrigation principles useful in determining irrigation practices. The practices to be recommended should be based on sound irrigation principles and should be designed specifically in accordance with prevailing soil, crop, climate, management and economic factors. Irrigation practices should not be merely copied from irrigation practices reported as successful elsewhere without carefully comparing all factors involved. A permanently successful irrigated agriculture and the efficient use of limited water supplies requires that the irrigation, fertilization and other cultural practices all be adjusted carefully to match local conditions.

1.4.8 DRAINAGE NEEDS AND METHODS

Agricultural drainage is the provision of a suitable system for the removal of excess irrigation/ rainfall water from the land surface so as to facilitate available soil moisture condition for the growth of plants to give sustained yield; to prevent soil moisture accumulation to allow earlier sowing. The raise of subsoil water level and development of salinity/alkalinity in soils are the causes of deterioration of irrigated lands. Therefore, drainage is provided in all irrigated lands especially in Tank and canal irrigation system.

The drainage is classified into two major systems: (i) surface drainage; and (ii) sub-surface drainage. The system of providing large outlet channel for the removal of excess water from excess rainfall/from over irrigation in surface or in below the surface depends on the character of soil and slope, etc. There are different systems of surface and sub-surface drainages adopted. In India, drainage is not considered important and hence the yields of crop are not upto the expected level. The research work is very limited especially in Tamil Nadu. More research work should be taken to increase the yield and save water in the coming years, to suit the conditions in the field.

1.5   WATER MANAGEMENT FOR IMPORTANT CROPS

1.5.1   RESEARCH FINDINGS

India is a vast country. The climate, rainfall, soils, seasons and the crops and cropping pattern are widely varying. Hence research is being carried out by many institutions/universities throughout India. The All India Coordinated Research on Water Management has permitted extensive research in various regions for various crops. The research findings on water requirement and scheduling of irrigation for important crops are summarized below:

1.5.1.1   Food Crops

1.5.1.1.1   Paddy

About 45% of the irrigation water is given to grow paddy crop in the country. The water used by the farmers in the project areas vary from 1500 to 2000 mm and therefore large quantity is wasted in addition the yield is also affected. Though the average production in advanced countries and even in countries like South Korea, Japan, Taiwan, Egypt is more than 7 to 8 tons/ha, the average yield in India is less than 3 tons/ha. By proper water management, it is possible to increase the average productivity more than 7 tons/ha.

The studies conducted on water research for paddy have indicated that there is no need to store more than 2 to 3 cm depth. In fact a thin film of water can saturate in the early stages of crop growth, i.e., 2 to 3 cm depth of water for about 30 days before and after heading time and again saturation in the last stage will be sufficient to get good yield. The experiment conducted to know the exact ET has indicated that the ET for paddy is only 600 to 800 mm for a crop duration of 100–140 days. Adopting this water saving method, nearly 30–40% of water could be saved without affecting the yield. The results of alternate drying and wetting conducted in many locations in India have revealed that there is no significant difference in yield compared to submergence. This method is being slowly adopted by the farmer when water availability in the canals/tank are limited. The reason for taking more yields by the well irrigated farmer is that he gives less water since it costs him more to give more depth of water.

1.5.1.1.2 Wheat

The irrigation requirements of wheat were investigated in many research farms in the last 10 years. At some Center, the response of irrigation schedule have been studied at varying levels of soil fertility, crop sequence, pre-sowing irrigation depth, weed control practices, etc. The results have revealed that on an average irrigation should be applied at IW/CPE ratio 0.90 and 6 cm of irrigation gave significantly higher yield. This required 4 to 6 irrigations depending on the seasonal rainfall and climate. The irrigation requirements vary from 25 to 35 cm. It was found that the reduction in yield occurred when the crop was irrigated at IW/CPE ratio 0.6.

1.5.1.1.3 Sugarcane

Sugarcane is fibrous rooted crop and most roots are active in first 60–90 cm depth of soil. For a maximum production, soil moisture should be maintained at a high level which will allow stalk elongation. Lot of research is being carried for water requirement, scheduling of irrigation and irrigation methods. Depending on climate and variety, the ET requirement of sugarcane varies from 1800 to 2500 mm. The highest yield was obtained when irrigation was applied at 20% depletion of available soil moisture. Irrigation should be given at IW/CPE ratio of 0.75.

From experimental research, it is found that frequency and depth of irrigation should vary with growth periods. During emergence and establishment period, light and frequent irrigations are preferred. Water deficit, during the establishment period and early vegetative period, has an adverse effect on yield as compared to water deficit in later growth periods. It was also found that during the ripening period a low soil moisture content is enough. The irrigation methods are furrow and skip furrow/paired row, drip and sprinkler systems. Recently Sustainable Sugarcane Initiative of irrigation (S.S.I.) method is followed using sub surface drip with fertigation to get higher yield (i.e., more than 100 T/acre).

1.5.1.1.4 Cotton

Depending on climate and length of growing period of the particular variety, cotton needs about 600 to 700 mm to meet its water requirement. From experimental results, it is found that the early vegetative period, crop water requirements are low or 10% of total. They are high during flowering period. Abrupt changes in water supply will adversely affect crop growth. Irrigation at 25% available soil moisture range was found to be best and the corresponding IW/CPE ratio was 0.6. Suitable irrigation methods are furrow, skip furrow/paired row, alternate furrow, sprinkler and drip system of irrigation.

1.5.1.1.5 Sorghum

For high production, crop water requirement of 110 to 130 days sorghum varies from 450 mm to 550 mm depending upon the climate. The crop coefficients (key) available from the research are: Initial stage 0.4, vegetative phase 0.7 to 0.75, flowering phase 1.0 to 1.15, late season stage 0.75 to 0.8 and at harvest 0.5 to 0.55. From the experimental results, it was found that the crop performance is best when irrigated at IW/CPE of 0.7 or irrigation at 50% soil moisture depletion. It is also found that when the crop has been well watered through the first two growth stages (vegetative and flowering), irrigation during grain filling period may be of no benefit to yields and may be even loss to irrigation water.

1.5.1.1.6 Maize

Maize is an efficient user of water in terms of total dry matter production and among cereals it is potentially highest yielding grain crop. For maximum production, a medium maturity grain crop requires between 400 to 500 mm of water depending on climate. Irrigation at 60% available soil moisture is better for getting higher yields. Optimal IW/CPE ratio was 0.9. It was found that maize subjected to moisture stress at silking and tasseling was most severely affected as far as grain yield was concerned.

1.5.1.1.7 Groundnut

Depending on climate, the water requirement ranges from 500 to 600 mm for the total growing period for groundnut. Irrigation at 40% available soil moisture (60% depletion) was found to be the best (irrigation of the crop at 0.6 IW/CPE ratio, depending upon climatic factors). The flowering and pod setting and pod filling periods are most sensitive to water deficit. Application of organic mulches like rice straw or gliricidia leaves to a depth of 2.5 cm saved 2 irrigations with 10% higher yield than from non-mulch field.

Check basin methods are being highly suitable for groundnut crop. Optimum bed size of 20 to 30 m² on light textured soils, sprinkler irrigation offers advantages by light and frequent water application.

1.5.1.1.8 Bajra

The water requirement varies from 350 to 450 mm. In one place, the grain yield was maximum at IW/CPE ratio of 0.50 and the grain yield increased with increase in the nitrogen dose from 0–105 kg N/ha. In another place the yield increased upto IW/CPE 0.75. The 4 cm irrigation depth was superior than 6 cm of irrigation depth.

1.5.1.2   Vegetables

The water requirement of tomato, brinjal and bhendi and other crops have been worked out. The irrigation is done at 50% ASM and IW/CPE = 0.6 to 0.8. The water requirement is about 600 to 750 mm depending on the duration and the season. The suitable method of irrigation is furrow, skip furrow/paired row and drip irrigation.

1.5.1.3   Fruit Crops

1.5.1.3.1   Banana, grapes, and papaya

The irrigation is given by the basins around the tree for most of the fruit crops. The size of the basins followed at present covers the entire area and equal to the spacing of the crops. For Anab-e-Shahi grape, the basin size is 8 m x 4 m and for the banana and papaya, it is 1.8 m x 1.8 m. From the experiments conducted in many years, it is found that the basin size can be reduced to 50% without affecting the yield, thereby large quantity of water could be saved.

1.5.2   PRACTICES IN WATER MANAGEMENT

Though various water saving devices for many crops have been worked out, still these methods are not widely used by the farmers. When there is water scarcity, the turn system is followed in paddy irrigation. Similarly, when there is drought and water supply is meager, farmers follow the irrigation scheduling as recommended by the scientists. Hence there is a long way to go for introducing all the research findings in the farmer's field, including drip irrigation.

1.6 ADVANCED METHODS OF IRRIGATION

1.6.1 STATUS OF IRRIGATION

Irrigation is as old as civilization in India. Though there are developments and improvements in all fields, practically there is no change in irrigation system/method. The farmers follow the same old practice which their great grand-fathers used to have 1000 years back. Experiments conducted to find out the optimum length of furrow and border length by various research workers and evolved many formulae. It was found that the water should reach the end of furrow/border within one-fourth the time the water takes to go to the root zone depth of the crop. Recent studies have indicated that by introducing alternative furrow or paired row irrigation for row crops, about 30–40% of water could be saved. Results are available about the size of the check basin to be formed which depends upon the soil, discharge rate, etc. Research was carried out to find out the optimum basin size for various fruit crops. These are some of the improvements made over the traditional surface irrigation methods.

In advanced countries, some advancement in irrigation has taken place when they introduce the sprinkler irrigation. This was used extensively after World War II, when aluminum pipes were invented. In 1960s, a breakthrough was done in irrigation especially in Israel by introducing drip irrigation. This method is now used by farmers in many countries in the world. In Israel sprinkler irrigation is being replaced by drip irrigation is used for all crops including banana, grapes, vegetables, cotton, fruit crops, etc.in more than 80 to 90% of the area.

1.6.2 SPRINKLER AND PERFOSPRAY IRRIGATION

In recent years, there has been a rapid increase in the use of sprinkler or over-head irrigation. This is a method of distributing water in pipes under pressure and spraying it into the air so that it breaks up into small water droplets and falls to the ground like natural rainfall. This need less water and labor than surface irrigation and can be adapted to more sandy and erodible soils on undulating ground.

Perfospray irrigation is an alternative to the sprinkler irrigation. They consist of pipes with small holes or nozzles along their length which water is sprayed under pressure. Small holes of 1 to 2 mm dia. are drilled into the top side of the pipe so that water sprays out in all directions wetting a rectangular area. The pressure requirement is less than the sprinkler irrigation. This method is practiced mainly in USA, Europe, Australia and Israel. In India too, this method is being adopted at various places for various crops especially in North India.

This method is very much suitable for lands of any topography especially undulated and sloping one. It can be used for all crops except rice and is ideally suited to the tea/coffee. It avoids wasteful surface runoff deep percolation below plant root zone. About 30–40% of water can be saved if this method is introduced.

Experiments were conducted in many places in the country for potato, wheat, maize, bajra, groundnut and cotton crops using sprinkler irrigation. The saving of water is about 25% to 40% and the yield increases will vary from 10 to 30%.

The sprinkler irrigation is replacing the surface irrigation in developed countries owing to labor requirement, higher efficiency of water use, ability to avoid frost attack, Possibility to apply fertilizer in solution, Further, land leveling is not necessary as in the case of surface irrigation. Through suitable extension technique the method can be popularized and more area can be brought under sprinkler irrigation in India, as suggested by Task Force on Micro irrigation. About 42.5 M.Ha are suitable for sprinkler irrigation for crops like cereals, oil seeds, pulses, tea, coffee, fodder, etc.

1.6.3   DRIP IRRIGATION

Drip irrigation is a relatively new method of crop irrigation with considerable water economy. In this system water is applied to plants through a network of tubings at a relatively low pressure. This irrigation concept was first introduced in Israel by Dr. Symcha Blass in 1959. At present it is widely practiced in USA, Australia, Israel, Hawaii, South Africa and South Europe. For India and other developing countries it is relatively a new method yet to find wider application. This method has many advantages especially in arid agricultural regions characterized by poor saline soil. The advantages of drip irrigation besides economy in water use are:

a. Saving of water
b. Saving of labor
c. Marked increase in crop yields
d. Possibility of using saline water
e. Shorter growing season
f. High quality produces
g. Versatility of undulating lands
h. Increased fertilizer use efficiency
i. Decreased tillage
j. Weed control cost savings

The disadvantages are that it is costly when compared to other methods and the plant growth under drip may develop shallow confined root zones since only a smaller volume of soil is wetted around the plant. Generally, in surface method less than half of water released reaches the plants. Of the water given the plants use only about 50% and remaining is lost by deep percolation and evaporation from the soil. Larger quantities of water need to be conveyed with additional storage facilities and channel capacities and with larger irrigation structures and extensive drainage arrangements. All these involve heavy capital investment. Further, about 5 to 10% of the cultivable land is occupied by the irrigation and drainage channels.

Drip irrigation provides small controlled amount of irrigation water at frequent intervals for the root system of plants. The drip system generally consists of: (i) A head unit connected to main water supply to the field which includes filter valves, water meter, pressure gauge and fertilizer apparatus; (ii) Main and sub-mains of suitable diameters according to the need generally 5 or 7.5 cm diameter PVC pipes; and (iii) Lateral or distribution pipes of small diameters 12 or 18 mm with nozzles or orifices.

A cheap drip irrigation system has been developed at Tamil Nadu Agricultural University, Coimbatore, India in accordance with the above description and it is much cheaper. This system works at low pressure and for emitters, one mm diameter holes with sockets are provided; corresponding to the location of each plant, high cost nozzles are avoided to reduce the cost. The optimum length of lateral is about 15 to 20 meters. The distance between the laterals depends upon the crop spacing. The system can be laid after the land is prepared for planting. The operation of the system is simple and involves only opening and closing of valves. At the end of season, the pipes can be gathered leaving the field free to the next crop. The water is given daily or once in two days depending upon the crop and climatic condition. The soil moisture level in wetted zone is always kept near the field capacity and hence crop growth is better in this method.

Research findings throughout the world has shown that trickle irrigation/drip irrigation is more suited to high value commercial crops, widely spaced vegetables, orange and other fruit trees, plantation crops, coconut and are canut trees, in the sandy area and slopy and undulated lands.

Experiments were conducted at Tamil Nadu Agricultural University, Coimbatore – Tamil Nadu for many vegetable crops, banana, sugarcane and cotton. The series of experiments conducted for variety of crops have indicated that the water used in drip method is only about 50% of control method (furrow or basin) and at the same time the yields are also increased by 10 to 40% in many cases, as the moisture content of soil is always near field capacity compared to conventional method. The germination in sugarcane and cotton was very good. It was also noticed that banana plants irrigated by this method flowered earlier than those in control. Further, the water is given only at the root zone to the required depth and hence the weed growth is also decreased by 50 to 60% in drip plots. The cost of the system vary from Rs. 8000 to 30,000 per acre (Rs. 20,000 to 75000 to 80000 per Ha) depending upon the spacing of crop (One US$ = 60.00 Rs.). The potential area for drip irrigation in India is about 27.0 M.Ha, but the present area is only about 2 M.Ha.

1.6.4 PITCHER IRRIGATION

In sandy areas, pot irrigation was given in some parts of South India by the farmers to save water. In this method, water is given to each plant by the farmer/worker once in two or three days according to the climatic conditions. Recently, scientists have designed a system in which mud pots are buried in the soil and by making small holes at bottom of the pot and inserting coconut fibers, the water is released drop by drop to the plants.

The experiments conducted with tapioca plants, fruit trees had proved that in this method it is possible to reduce the crop water requirement. This is called Pitcher method of irrigation. It can be introduced especially in large spaced fruit and orchard crops and tree farming.

1.6.5   ADOPTION OF IRRIGATION METHODS

The improved method in surface irrigation is followed by the farmers during drought years when there is great demand of water. The advanced method of irrigation like sprinkler and drip are extensively used in advanced countries, but not in India since the initial investment is prohibitive to the farmer. Since water saving is substantial, the Central and State Governments are now encouraging the farmers to introduce this system by giving subsidy to the tune of 50% to 75% for sprinkler in the north and drip in the south India. Hence, it is catching up now and large areas will be brought under this system of irrigation in next 10 to 15 years.

1.7   CASE STUDIES AND LARGE ADAPTATION

1.7.1   WATER TECHNOLOGY CENTERS IN INDIA

a. General

Water is the resource needed for various activities especially for agriculture –irrigation. The demand of water is increasing day by day for the growing population. Water is considered as a liquid gold in Tamil Nadu, India since the available water in Tamil Nadu is only 650 M3/p/y if all the available water is taken into account and it will be reduced in the coming years due to population increase. It is a valuable natural resource not only to human being but also for the entire eco system for the sustenance and development.

b. Establishment of W.T.C./W.M. Institutions in India

The 1st W.T.C. was established at the I.A.R.I. New Delhi as an inter-disciplinary facility for research, teaching, training and extension in water management. It was established in 1969 with technical collaboration with the University of California, Davis and partial support from the Ford Foundation U.S.A. The center has evolved into a unique institution addressing a wide range of activities centering around W.M.at all levels. Knowing the need and importance of water due to the scare condition of Tamil Nadu the author has prepared a project report to establish a W.T.C. at T.N.A.U., Coimbatore and submitted to the S.I.D.A., Sweden to get funds during 1980–1981 through Govt. of India. SIDA approved the proposal and contributed a substantial fund to establish the center at Coimbatore in 1982 to play a lead role in developing appropriate Technology, Knowledge base of the available water and maximizing agricultural production and evolve suitable water management for different crops in different agro climatic zones and training. Subsequently the Govt. of India and State Agric. Universities have established research centers on water in various states to cater the needs of the country. At the same time, all the States have established Training Institutions for Water resources and management including drip and sprinkler irrigation system.

c. Mission and Vision

The major thrust areas for research in WTC in New Delhi are: Development of farm level technologies for water management. With the initiation the command area development program in 1974. The research in the center was expanded to include development of scientific guidelines for efficient water distribution and management in irrigation commands. The studies will find solution to W.M. in northern part of India.

The WTC at Coimbatore, has concentrated the following research projects to solve water problems facing South India:

- Cropping System management.
- Irrigation and drainage.
- Micro irrigation and fertigation.
- Watershed development and management.
- Water shortage (stress) management.
- Water quality and water recycling.

The idea is to bring at least 4/5 WTC throughout the country to cater the needs of the entire country. Subsequently many WTC & Water Management Institutes were established throughout India.

d. Major Findings/Technologies Developed in Water Management Institutions

i) WTC–IARI, New Delhi

This center has developed state of art Lab facilities for carrying out inter disciplinary research in agricultural water management, modern tools like Remote Sensing, GIS and simulation modeling techniques used for developing soft were information system, design support system (DSS) Expert system, etc. for enhancing water use efficiency at all levels

Fruitful and mutually benefitting knowledge and experience were shared with majar irrigation command areas like Nagarjunasagar, Sharda Sahayak, Mahi, etc., besides large number of farmers, tea plantations and river valley projects.

The precision farming development center (PFDC) for the semi-arid regions of India was established at the WTC. The center based on its field research, developed several technologies, including hardware's and computer software for estimating water requirements, modification of crop geometry, use of mulches, cost reduction, fertigation, sub surface drip system and use of saline water and domestic sewage in vegetable crops through micro irrigation.

ii) WTC, TNAU, Coimbatore

The major findings/technologies developed at WTC, Coimbatore are as follows:

- Irrigation methods and scheduling for various field crops have been developed based on soil, water and agro climatic conditions of the various regions of the state.
- Development of efficient water management practices include cyclic submergence of water 1–3 days after disappearance in rice crop and alternate furrows, paired rows, broad bed furrow of irrigation in dry crops.
- Gradual widening of furor in sugarcane and graduate widening of basins in banana were efficient in water use.
- SRI method for paddy cultivation and SSI method for sugarcane crops were developed to reduce water requirements and to increase the yield of these two crops.

- Mulching with coir pith and polythene film for soil moisture conservation was accomplished.
- Drip irrigation to wide spaced crops like coconut, grapes, orchard tree crops and close spaced crops like sugarcane, banana, tapioca and vegetable were accomplished wherein water saving, yield and quality increase were demonstrated.
- Affordable micro irrigation for kitchen garden, small and marginal farmers, etc.

1.7.2 EXTENSION ACTIVITIES IN WATER MANAGEMENT

Information regarding water requirement of various crops (when to irrigate and how much to irrigate) is available since scientists have been working for the same for the last 25 to 30 years. Similarly many firms in India are fabricating and selling the equipments and materials for sprinkler and drip irrigation. Still this technology has not been popular and has not been taken up by the farmers on large scale. The reason for not introducing the better water management practices is that this information and findings have not gone to the farmers. The extension agencies available at present are not aware of the technology themselves and they are not given the responsibility for popularizing the same. However, whenever and wherever there is scarcity of water, the farmers approach research institutions to get the latest technology and adopt them in their fields if it is suited.

In order to popularize the technique, the research scientists working in various projects need to take large scale demonstrations or operational research projects in farmer's field in order to popularize the same among the farming community.

1.7.3 OPERATIONAL RESEARCH PROJECTS (ORP)

In the coordinated research project on water management, large scale demonstration of the research findings are made in the farmer's fields. The Tamil Nadu Agricultural University has taken up this ORP in the major irrigation projects namely Lower Bhavani Project, Periyar – Vaigai Project and Thanjavur delta. An area not less than 100 to 200 acres are selected in a branch canal/sluice and improved water management practices are introduced including the appropriate agricultural technology in growing crops. An equal area is taken adjacent to the demonstration plot for comparative study. The findings of the studies in the last 2 to 3 years have revealed that by following the methods as suggested by the scientists, they were able to save about 25–40% of water without affecting the yields, but the water has to be controlled and irrigated accordingly, whereas in the normal case (existing method) water is not controlled. By giving the water as recommended by the scientists by little land preparation, not only the saving of water by about 30–40% but also substantial increase of yield was noticed. Therefore, it is necessary to go for the water management practices for the entire area of the irrigation projects in the country.

1.7.4   CASE STUDIES IN IRRIGATION PRACTICES

1.7.4.1   Periyar–Vaigai Projects

Studies were conducted to find out the equity and productivity in different locations of the system and also in different areas. It is observed that when the farmers are confident of getting the required quantity of water, the water distribution is simple and all the fields get equal quantity of water irrespective of head or tail reach. It is important to note that the productivity is also more or less same throughout the canal command area. But this picture is reverse in an area where the farmers are not confident of getting enough water for the crop, i.e., in the tail end of the main canal. Because of this, the productivity is also varying very much. Therefore, in order to distribute the canal water without any problem, the first requisite is quantity if water management practice can be introduced in order to get more production with the available water. In many cases, the traditional crop and cropping patterns are followed. This has to be changed and further the cropping period should be changed to utilize the rain water more affectively. It is noticed that the ground watertable is almost at surface which affect the growth of crops and thereby the yield. Incentives should be given to use the ground water in a conjunctive manner to solve drainage problem and to pave way to increase the yield. In many cases water is supplied in the canal for longer period than the actual duration need of the crop. This is because the gap between the first and last transplantation in the same canal is very wide (i.e., 30–50 days). There is scope to save large quantity by proper planning and inducing all the farmers to take up land preparation and transplantation of the paddy within a short time. Of course sufficient quantity of water is to be given to achieve this. In a study, it is found that there is scope to take one more crop by proper planning and utilizing the residual moisture in the paddy field. For all the above it is necessary that the farmer and the water authorities should have more participation and understanding and required training should be given to all in water management practices.

1.7.4.2 Parambikulam – Aliyar Project – Old Channels

Studies conducted in Perianai channel of Aliyar basin of PAP have indicated that these channels were constructed long back and are not in good condition. The cross section has increased and the slope is not maintained. The seepage losses are very high. The pipe sluices are also not in accordance with the area under its command. If better water management practices are to be followed, modernization of the canal and sluices are a must. It is not even possible to close the sluices, or to close the sluices for controlling irrigation water. Under these circumstances it is difficult to introduce water management in these old syacuts. In addition, the irrigation is done from field to field only. There is great scope to reduce the water use in the syacut and the farmers are very much enthusiastic about that. The farmers are willing to go for Warabandhi system. But this could not be practiced until the system is rectified to carry and control the water supply.

1.7.4.3 Tank Irrigation in Chengalpat District

Studies were taken up to introduce the water management practices in tank irrigation. It is found that sluices are not maintained and hence the operation of the sluices to release the required quantity is a problem. The cross section of the channels has been reduced by encroachment and further it is badly silted up. None is bothered to maintain the channels and all the adjacent fields are flooded. There is no one responsible to allocate water in different sluices, based on the area of crop and stages of crop. The gap in transplanting of the crop in the project is about 30 to 45 days and this adds to the wastage of water. The seepage loss is more than 50% and the time allotted to tail enders is not sufficient to irrigate their land. Though there is possibility of growing other than paddy especially in the first season, many farmers cannot change it due to overflow of water from field to field. In order to introduce the technology in water management, the following arrangements may be made:

a. Tank bund, sluices, channels should be maintained as per the norms.
b. Main channels may be lined.
c. Sluices for medium/small tanks should be opened only during day time.
d. Discharge should be based on the water requirement of crops.
e. Land preparation and transplanting must be completed within 10–15 days.
f. There should be sluice committee to operate and regulate the supply. The farmers who have land under the sluice should select their chairmen and members.
g. On Farm Development works should be taken up for all tanks.
h. Incentives should be given to the farmers to use ground water which will not only save water, but also help to increase the yield.
i. The officers in charge of the tank and the farmers should be given training in water management practices.

1.8   ALLOCATION OF LIMITED WATER FOR IRRIGATION

1.8.1   MAXIMIZATION OF WATER UTILIZATION

In many areas of the world, the available or even the potential water supply is inadequate to meet industrial and domestic uses and also to satisfy the irrigation requirements of desired cropping patterns on all available land. The need to provide water for power generation, navigation, waste disposal, etc., must also be considered. This means that consideration should be given programs which can maximize the utilization of all water available in an area by allocating water supplies among conflicting uses and areas of use so as to achieve the greatest public benefits. Where the agricultural objective is to maximize profits, this will favor maximizing production per unit of resource input-water, land, labor or capital, depending upon which one is in the shortest supply. But in under-developed countries facing food shortages, where the national and economic objectives will likely require rapid increases in agricultural production, means should be sought to maximize total food production per unit of available water supplies.

For optimum crop production, one of the first requisites is to ensure that required water supply is available to the crops in the right quantities and at the right times. This is essential for favorable crop production. Efficient use of water and also for providing the assurance and incentive farmers need to adopt suitable cropping patterns and undertake the intensive production practices required for successful irrigated agriculture.

Water allocation and management programs pose serious problems for Government policy makers and administrators.

On a national level, the question arises "How should the total available water resources which can be developed in various projects be distributed between kinds of users and areas of use in the country?" At the level of irrigation project planning, the question arises "What crops are most suitable from the standpoint of domestic and export needs and what are their water requirements?" Information on these questions as well as on many others including suitability of topography, soils, etc. will determine how big an area can be irrigated by the water available in given projects. On the level of the individual water user, he must decide what crops to plant and how much area is to be devoted to each crop.

To allocate water efficiently for use in crop production, it is necessary to know the relation between water supply and the yields of various crops under given climatic conditions and given levels of other essential inputs such as fertilizer. These production functions enable project planners and farmers to predict probable reductions in crop yield corresponding to given reductions in water allocated. This information can be used to determine how much area of what kind of crops may be optimal to plant with given supplies of water in storage or forecast. Since available water supplies within given projects often vary from year to year, information on the production functions of different crops will allow rational decisions on what portion of the irrigation supply to use for perennial and other valuable crops. This information can be used for perennial and other valuable crops. This information would also allow the planting of additional areas with predictable probabilities of given yield reductions under stated water supply conditions. Such analyzes would also indicate the desirability of planting crops which are less affected by reduced water supplies or have lower value on areas which could not be guaranteed a normal water supply for full production. Further, the planner and the farmer would have the basis for determining what areas should be left unirrigated in given seasons.

Ultimately, production functions should permit production for a particular land situation. Given its climatic environment, what is the expected physical increment in production and economic return would be for selected crops? Such information is essential both for evaluating the potential or irrigation development and forecasting economic return or capacity to repay costs of the irrigation development.

1.8.2   RESEARCH ON PRODUCTION FUNCTIONS IN OTHER COUNTRIES

Research on water production functions were conducted up by Kleinman and Heady, Steward and et al. in USA. They have developed production functions for principal crops including alfalfa and. wheat. Grimes and et al. have developed yield with the combination of water and nitrogen for cotton. It is found that increasing amounts of nitrogen are required to maintain yield as water supply declines. Conversely at lower levels of nitrogen fertilization, more water is needed to achieve an equal yield. Though it is difficult to obtain useful crop yield versus water use function, because yields are seldom limited by water stress acting alone. However, it is possible to synthesize useful yield-water use function for some important crops from the large amount of published field research data.

Scientists in Israel are working on the response curves for water and fertilizer; and based on the results recommend the allocation of water to the farmers for various crops. The very fact, that they are able to increase the production in spite of reduction in allocation of water for agriculture, will indicate that it is possible to extend the same to all countries and for all crops.

1.8.3 RESEARCH ON PRODUCTION FUNCTIONS IN INDIA

Much work has not been done on the vital areas in India. Studies have been taken up at Water Technology Center, Tamil Nadu Agricultural University Coimbatore on water fertilizer production for crops like cotton, maize, ragi and jowar. Studies on paddy are being carried out at many research stations. The findings will go a long way to allocate water and fertilizer during scarcity and abundant situations in the country. The main objective in all these studies is to get maximum production from unit quantity of water and fertilizer.

1.9   SOCIAL, ECONOMIC AND LEGAL CONSIDERATIONS

1.9.1   SOCIAL CONSIDERATIONS

Since there was not much demand for water in the past, abundant quantity was diverted for irrigation. The duty in some cases works out less than 30 acres/cusecs. Scientists have provided not only in the research stations, but also in large scale demonstration plots that the allocation of water in these channels is more. But they are demanding for their right. Not the right to take only the required quantity of water, but the amount of water they used to get. Further, it has further complicated since there are no water management policies, uncertainties of water supply and numerous Governmental entities involved in planning for and administration of water resources.

The scientists have convinced that by proper water management practices, large quantity of water could be saved and the same can be diverted from one basin to other basin if necessary, but this could not be done for various social and political reasons. This is true not only in the country, but also in many countries especially in the developing countries. Therefore, any water management development and improvement should take note of the social aspects of the area before talking up for implementation.

1.9.2 ECONOMIC CONSIDERATIONS

Most of the irrigation projects were constructed after independence of India. The area of irrigation has increased from 20 M.Ha to more than 100 M.Ha in the last 30 years. In preparing the plan and estimation for the irrigation projects, much attention was not given to the distribution of water after the sluice/pipe outlet level and removal of excess water and rain water from the field. One reason for this is that the cost per acre may go up if the expenditures for these items are added in the estimate since ceiling is fixed to get the estimate sanctioned per acre irrigated for a new/old project. However, it is now found that since facility is not available to distribute the water as per the requirements of water and to dispose of the excess water, large quantity of water is wasted which affects the land and production. In order to use water efficient studies have revealed that necessary control structures, separate irrigation/drainage channel are necessary. To provide all these facilities, on farm development works are to be carried out in all the command area. The old irrigation systems in some cases should be completely modified taking into account the peak water use of the crops in addition to construction of sluices, channels and other control structures to manage water distribution.

In all irrigated fields, the farmers have constructed the wells and they have not provided proper conveyance structures to carry water. The well water is costly compared to canal water. Much attention should be paid in the use of water. Advanced methods of irrigation like sprinkler and drip irrigation can be introduced in the well-irrigated farms. This will require about Rs. 20,000 to 70,000/Ha (One US$ = 60.00 Rs. in this chapter).

Since most of the farmers are poor and cannot afford to take up all the above-mentioned improvements, Government and Banks should come forward to take up all the developmental activities for the national interest. In fact even in advanced countries like USA, Israel, Greece, Government is helping the farmers by subsidy or reduction wherever water is saved by the farmers.

The advent of modern computers of the development of systems analysis for water resources problem has provided a powerful tool for dealing with the technical as well as social, economic and legal constraints in a complex water system and for ordering alternative solutions so that decision makers have the opportunity to select among alternatives knowing their relative costs and consequences.

1.9.3 LEGAL CONSIDERATIONS

To introduce water management practices and increase the water availability in an area, there are many legal barriers and considerations. The concept of 'Riparian Right' is coming in the way of, for introducing better water management in many irrigated areas. The legal aspects of weather modifications are being studied by legal experts since it is questioned by neighboring farmers with reference to judicial statutes. Administrative regulations as well as technical reports, to determine what legal norms should be in the future be applied to local, national and international consequences of weather modification activities. The other areas of research which may well lead to social economic and legal problem when man seeks to exploit has new research on watershed management and runoff control by soil and vegetative management, ground water recharge in special areas. Regulation of surface and ground water levels between different uses and areas of use and the use of more efficient irrigation practice is necessary, which may in turn reduce return flow which other users may depend. All these aspects are to be considered while implementing water management practices in the command areas.

1.10 FUTURE FOCUS ON RESEARCH, TRAINING AND DEVELOPMENT: WATER MANAGEMENT

1.10.1 UTILIZATION OF EXISTING KNOWLEDGE

Research has already provided considerable knowledge on water management, storage, conveyance, application of irrigation water and water use by crops. Substantial improvements in water use efficiency could be achieved by utilizing knowledge currently available. This calls for programs of education both for farmers and for personnel of water supply agencies and to extension services. A word of caution, however, is desirable. Program planners have too often thought that the solution of water problems called merely for the use of practices found to be successful in other locations. Actually many water management practices will require adaptation if they are to be fully successful in given situations.

1.10.2   NEED FOR ADAPTIVE RESEARCH

Such research is needed to refine information existing in more developed areas and to adapt it to conditions in the area of proposed use. Doubtless substantial improvements in water management practices and in agricultural production could be achieved by suitable adaptation of knowledge existing in other areas. This calls for experienced research personnel capable of utilizing effectively the information developed in other situations and adapting it to the problems of the area concerned. Contrary to opinions expressed by some, this type of research should not be looked upon as a second classes research activity. Rather it is a type of research which calls for great experience and ability to analyze and interpret the interplay of local factors which control the application and usefulness of information from other areas. Persons involved in such research should not be considered to be carrying on research of less importance than those engaged in what is usually called fundamental research.

1.10.3   NEED FOR NEW RESEARCH INFORMATION

As pointed out above, increasing world population and rising per capita water use creates/ needs for new information to solve old problems never well resolved and now increasingly demanding solution together with new problems brought on by man's increasing density and intensity of water use. Personnel engaged in water research need to have solid scientific and engineering backgrounds so that basic knowledge can be built on to solve practical water problems. Emphasis should be placed on those research programs which will provide solutions to water problems not just by refining today's technology, but through developing entirely new technologies.

1.10.4 DIRECTION FOR FUTURE RESEARCH

The amount spent and provided on the Five year plans under irrigation is huge in India. Already about 100 billion Rs. have been spent and in the next 20 years the amount required may be in the order of 1×10^6 to 2×10^6 million Rs. Compared to the dimension of the problem, the attention paid on the vital subject is very meager. Research on water management should be intensified for canal, tank and well-irrigated areas. The work done so far relates to the water requirements of crops in the research stations, but not on a comprehensive manner for the irrigation project. Studies should be taken up to evaluate the entire project and monitor and allocate the water for optimum production. This can be done with the cooperation of irrigation department and farmers. Further, research data on how the available water could be allocated to get maximum profit by the farmers are not available. What are the alternatives and how conjunctive use of surface, ground water and rain water may be achieved. Work on the drainage, methods of drainage needs more attention in the years to come. A cheap and trouble free drip irrigation should be designed and introduced. Yet another area of research is about the cheap method of construction wells, lifting the water at less energy, conveying the water with maximum efficiency and application of water to give maximum production and profit from unit quantity of water. All the scientists working in the field of water management should dedicate themselves to find solutions as water is going to be scarce commodity and may cost more in the years to come.

In addition, large scale demonstration/ORP should be taken up in well, tank and canal irrigated area in order to popularize the technique. The farmers and extension agencies should be involved in all these activities so that the knowledge gained should be used without any waste of time.

1.10.5 RESEARCH NEEDS AND ACTION PLAN TO MITIGATE WATER PROBLEM

International Water Management Institute, Colombo – Sri Lanka has predicted that there would be an acute water storage worldwide by the year 2025. About 33% of world population comprising about 45 nations will be most severely affected by this problem. India is one of these nations. It also warned that most of these countries will have to import a substantial portion of their cereals consumption, if they do not develop at least 25% more new primary water resources. This is true in the case of Tamil Nadu as the supply demand gap of water is estimated as 24% in 2020. Further we are experiencing the water shortage from time to time. In order to combat water problems for long term, an integrated effort must be made by the Government and people from all walks of life. The action plan for tackling the water problem of the state for the future research agenda are detailed in the following sub-sections:

1.10.5.1   Area of Future Research

To satisfy the demand of water, research should be done in an integrated way. The suggested items of research and development works are as follows:

a. More crop per drop (water).
b. Theme oriented research – attacking the problem with different disciplines joined together.
c. Climate change impact on Agriculture – study by Meteorology/ Energy/Engineering/Breeder/Water Department/joined together.
d. Increasing productivity in all crops using the available technology by holisticapproach – Time bound R & D.
e. Value added products by farmers. How to go about for different crops – R & D.
f. Marketing help to farmers for all crops to have incentive to increase production.
g. Use of farm implements/machinery by all farmers through awareness and workshops and advisory committee and preparation of a realistic plan of approach.
h. Preparation of water vision of Tamil Nadu.
i. Preparation of an action plan for drip irrigation and sprinkler irrigation to cover about 20% to 25% of the irrigated area in 2020/2025.
j. Crop/cropping patter changes – PAP and other irrigation projects depending on water availability and especially in Tank irrigated areas where paddy yields are very low.

The areas of activities to mitigate water problem are:
- Soil water plant relationship.
- Quality of irrigation water and management, reclamation of polluted water and land.
- Watershed management.
- Rain water management (urban/rural).
- CAD/OFD – Command Area Development and On Farm Development.
- Farmer's participation.
- Impact assessment of irrigation projects especially in the old irrigation projects.
- Contingency plan – drought/flood.
- Management of rural industry effluent water and municipal sewage water – Reclaiming for reuse.
- Irrigation economy. Adoption of proven technology in water management and introducing advanced method of irrigation like sprinkler and drip on large scale.
- Irrigation and drainage Engineering.
- Collection and Supply of data for research and planning.
- Advisory.
- Publications.
- Teaching.
- Training.
- Seminars/workshops in villages/towns.
- Demonstrations pilot projects.
- Operational research projects.
- Soil conservation, land utilization and reclamation.

1.10.6 TRAINING NEEDS IN WATER MANAGEMENT

Water Management is very new branch of science and hence there is no cadre staff to tackle the problems. Each Department/organization is having its own professional staff to tackle the problems and extension activities. Sufficient qualified persons in the field of water management are not available even for research organization and teaching positions in the Universities. In spite of this, Research and little Extension works are carried out in a small way. Trainings are provided in Agricultural and Engineering Universities and research institutions, to the farmers and extension staff. The demand for more staff will be there since large area is going to be brought under irrigation in the next 15–20 years.

Obviously to provide the guidance, services and training the Government need well trained staff. As mentioned such staff is scarce in the extension services in the Department of Agriculture, Agricultural Engineering and Irrigation. The engineers in Irrigation Department are specialized in civil engineering designs and their experience in designing irrigation aspects in farms are limited. Therefore, intensive training in Irrigation, Agriculture and related officers working with farmers as well as Engineers is indispensable. Based on the above facts all the states and central Governments have established number of water management training institutions with the help of USAID, World Bank, SIDA, etc.

The line of action thus provided is usually of short duration and should be considered for short term and partial remedy only. To provide necessary staff, it will be necessary that universities should include or reinforce curricula and courses in their regular programs which are geared to Agricultural water and Soil management.

Appendix A shows a list of WTC in India. Appendix B indicates 15 examples of water management practices in India (Figures A1.1–A1.15).

1.11 SUMMARY

The water supply on our planet earth is essentially fixed. But the demand is increasing day by day not only for agriculture, but also for other uses namely industrial and municipal use. Man can alter to some extent the special and time distribution of the water. But this will not be sufficient for growing needs of the community.

Fortunately some water management problems can be resolved at relatively low cost by utilizing existing knowledge. These opportunities exist in both the developing and the developed countries. Some important water problems can be dealt with by adaptive research which develops from existing information solutions fitted to the specific problems. Still other water problems will require new research to achieve desirable solutions. Man has major opportunities to augment water supplies through improved watershed management practices, more efficient measures for storage and conveyance of water and possibly by weather modification. Through research and use of improved water control equipment, irrigation efficiencies could be substantially increased in many areas.. Careful studies of the water-soil-plant- atmosphere system to establish optimum irrigation practices for specific crops can lead to substantial water savings and increased crop yields. Predication of crop responses to irrigation as provided by studies on production functions can provide a sound basis for water allocation between agriculture and other users as well as within agriculture also.

Sprinkler and drip irrigation can be practiced in suitable areas to save water especially in shallow and sandy tracts. There is great demand/scope for these systems of irrigation. Government is also interested in expanding this advanced method of irrigation.

Modern analysis techniques to both the water quantity and quality aspects of complex water resources system will permit the systematic arranging of alternatives and provide opportunities to evaluate costs of possible optimal choice which may be suggested by social, legal, economic experts/consultants.

KEYWORDS

- absolute scarcity
- arid region
- conjunctive use
- cultural practices
- desalination
- diversification
- evaporation
- germination
- harnessable

- hydrological cycle
- industrial waste water
- interception
- isolated manner
- legal barriers
- methodology
- on farm development
- operational research
- operational research project (ORP)
- paired row irrigation
- percolation
- perfospray
- production function
- project evaluation
- rechargeable
- reclamation
- regulating structures
- response curve
- sub surface drainage
- techniques
- turn system
- warabandhi system
- weather modification

REFERENCES

1. Dieleman, P. J. (1982). Farm Water Management as a global problem. FAO, Rome.
2. Goldberg, D. (1976). Drip Irrigation. Drip Irrigation Scientific Publication, Israel.
3. Megh R. Goyal (2015). Research Advances in Sustainable Micro Irrigation, Volumes 1 to 10. Apple Academic Press, Inc. Oakville, Ontario, L6L0A2 Canada.
4. Megh R. Goyal (2016). Innovations and Challenges in Micro Irrigation, volumes 1 to 5. Apple Academic Press, Inc. Oakville, Ontario, L6L0A2 Canada.

5. Megh R. Goyal (2016). Innovations in Agricultural and Biological Engineering, volumes 1 to 7. Apple Academic Press, Inc. Oakville, Ontario, L6L0A2 Canada.
6. Hagen, R. M. (1968). Pioneering research in water management. Paper presented at AAAS Symposium on New frontiers of Agricultural Research, Dallas, TX.
7. Johnson, S. M. (1978). Improving irrigation. Water Management, 15(3).
8. Melvyn, Kay. Sprinkler Irrigation Equipment and Practice.
9. Niranjan, Pant. Productivity and Equity in Irrigation Systems.
10. Rajput, R. K. (1982). Research on water management, progress reports of coordinated project for research on water management. ICAR, New Delhi, 1967–1982.
11. Sivanappan, R. K. (1982). Management of available water resources. Paper presented at the International symposium on polders of the world, Netherlands.
12. Sivanappan, R. K. (1980). Sprinkler and drip Irrigation. Paper Presented in the National Workshop on Water Management at Hanoi.
13. Sivanappan, R. K. (1977). Well irrigation for small and marginal farms. Paper presented at the XV Annual Convention of ISAE, Pune.
14. Sivanappan, R. K. (1975). Economizing the use of water by advanced irrigation practices. Proceedings of the Second World Congress, New Delhi.
15. Sivanappan, R. K. (1984). Recent Development in Water Management of Crops. Mimeograph.
16. Sivanappan, R. K. (1994). Drip irrigation in India, INCID, Govt. of India, New Delhi.
17. Sivanappan, R. K. (1998). Sprinkler Irrigation in India, INCID, G.O.I., New Delhi.
18. Sivanappan, R. K. (2008). Water Management for Sustainable Agriculture, TNAU, Coimbatore.
19. Sivanappan, R. K. (2002). Soil and Water Conservation and Water Harvesting. SIDA, Through Forest Department, Govt of Tamil Nadu, Chennai.
20. Sivanappan, R. K. (2005). Irrigation in 21st Century—Challenges and Solutions, TNAU, Coimbatore.
21. Sivanappan, R. K. (2012). Tamil Nadu Water Vision and Interlinking of Indian Rivers: Need and Importance. T.N.A.U., Coimbatore.
22. Sivanappan, R. K. (2013). Hi Tech Precision Farming in Agriculture and Horticulture Crops. Kissan World, 40(1), Chennai.

APPENDIX A:
LIST OF WATER TECHNOLOGY CENTERS AND CENTER FOR WATER RESOURCES AND MANAGEMENT RESEARCH IN INDIA

S. No.	Name of Center	City, State
1	Water Technology Center, IARI	New Delhi
2	Water Technology Center	Coimbatore, Tamil Nadu
3	Water Technology Center	Bhubaneswar, Orissa
4	Water Technology Center	Hyderabad, Andhra Pradesh
5	Center for Water Resources Development and Management	Calicut, Kerala
6	Center for Water Resources and Management, Anna University	Chennai, Tamil Nadu
7	Institute of Water Studies, Water Resources Organization	Chennai, Tamil Nadu
8	Madras Institute of Development Studies	Chennai, Tamil Nadu
9	Center for Water Resources, Karunya University	Coimbatore, Tamil Nadu
10	Water and Energy Research and Training center	Vivekanthapuram, Coimbatore, Tamil Nadu
11	Water management Forum, Institution of Engineers (India)	Ahmedabad, Gujarat State
12	Soil and Water Conservation Research and Training Center	Dehradun, U.P. State

APPENDIX B:
EXAMPLES OF PROJECTS ON WATER MANAGEMENT ENGINEERING IN AGRICULTURE.

FIGURE A1.1   Drip irrigation in rice.

FIGURE A1.2   Drip fertigation in tomato.

FIGURE A1.3   Drip irrigation in okra.

FIGURE A1.4   Drip irrigation in onion.

FIGURE A1.5   Micro sprinkler irrigation in groundnut.

FIGURE A1.6   Surface irrigation – broad bed furrow.

FIGURE A1.7   Drip irrigation in banana.

FIGURE A1.8   Drip irrigation in coconut.

FIGURE A1.9   Drip irrigation in cotton.

FIGURE A1.10   Drip irrigation in vegetables.

FIGURE A1.11   Drip irrigation for fruit trees.

FIGURE A1.12   Check dam.

FIGURE A1.13   Farm pond.

FIGURE A1.14   Farm pond for dry land.

FIGURE A1.15   Sprinkler irrigation.

CHAPTER 2

RAIN WATER HARVESTING: STATUS, PROSPECTS, CONSERVATION AND MANAGEMENT

R. K. SIVANAPPAN

CONTENTS

2.1   INTRODUCTION

Water is essential for all life and is used in many different ways. It is also a part of the larger ecosystem in which the reproduction of the bio diversity

Part of this chapter was presented at a local seminar by the author: *R. K. Sivanappan, 2006. Rain Water Harvesting, Conservation and Management Strategies for Urban and Rural Sectors. Pages 1 to 5, In: Proceedings of National Seminar on Rainwater Harvesting and Water Management 11–12 Nov. 2006, Nagpur. http://portal.unesco.org/geography/en/files/6192/11690988831Accepted_Papers_-_1. pdf/Accepted%2BPapers%2B-%2B1.pdf.*

depends. Fresh water scarcity is not limited to the arid climate regions only, but also to areas with good supply (rain). The access of safe water is becoming a critical problem. Lack of water is caused by low water storage capacity, low infiltration, larger inter annual and annual fluctuations of precipitation and high evaporation demand.

The term water harvesting was probably used first by Geddes of the University of Sydney, who defined it as "*the collection and storage of any form of water, either runoff or creek flow, for irrigation use.*" Several modifications of the definition have broadened the term to mean "the process of collecting natural precipitation from prepared watersheds for beneficial use." Myers at USDA – USA has defined it as the practice of collecting water from an area treated to increase runoff from rainfall. Recently, Currier of USA has defined it as the process of collecting natural precipitation from prepared watershed for beneficial use. Now a days water harvesting has become a general term for collecting and storing precipitation of runoff or creek flow, resulting from rainfall in soil profile and reservoirs both over surface and under surface. Previously this was used for arid and semi arid areas, but recently its use has been extended to sub humid/humid region too. In India, water harvesting means utilizing the erratic monsoon rain for raising good crops in dry tracks and conserve the excess runoff water for drinking and for recharging purposes.

This chapter includes general discussion on status, history, prospects, conservation and management of rain water harvesting under Indian conditions. The concepts in this chapter are also applicable to other countries.

2.2   HISTORY OF RAIN WATER HARVESTING

Although the term water harvesting is relatively new, yet the practice is ancient. It has been practiced as early as 4500 B.C. by the people of Ur and also latest by the Nabateans and other people of the Middle East. Although the early water harvesting techniques used natural materials, yet 20th century technology has made it possible to use artificial means for increasing runoff from precipitation.

Evenari and his colleagues of Israel have implemented water harvesting system in the Negve desert. The system involved clearing hill sides to smooth the soil and increase runoff and then building contour ditches to collect the water and carry it to low lying fields where the water is used to irrigate crops. By the time of the Roman Empire, these runoff farms had evolved into relatively sophisticated systems.

The next significant development was the construction of road catchments as described by the public works Department of Western Australia. They are so called because the soil is graded into ditches. These ditches convey the collected water to a storage reservoir. Lauritzan, USA has done pioneering work in evaluating plastic and artificial rubber membranes for the construction of catchments and reservoirs during 1950's. In 1959, Myers of Water Conservation Laboratory, USDA – USA began to investigate materials that caused soil to become hydrophobic or water repellent. Then he gradually expanded to include sprayable asphaltic compounds, plastic and metal films bounded to the soil compaction and dispersion and asphalt fiber glass membranes. During early 1960, research programs in water harvesting were initiated in Israel by Hillal and at the University of Arizona – USA by Cluff. Hillal's work was related primarily to soil smoothing and runoff farming. Cluff conducted a considerable amount of work on the use of soil sealing with sodium salt and on gravel covered with plastic membranes.

The research in India on this subject is of recent one. Work has been taken up at ICRISAT, Hyderabad; Arid Zone Research Institute, Jodhpur; Central Research Institute for dryland Agriculture (CRIDA), Hyderabad; Agricultural universities and other dry land research centers throughout India.

In Pakistan, in the mountainous and dry province of Balukhistan, bunds are constructed across the slopes to force the runoff to infiltrate. In China, with its vast population is actively promoting rain and stream water harvesting. One very old but still common flood diversion technique is called 'Warping' (harvesting water as well as sediment).

When water harvesting techniques are used for runoff farming, the storage reservoir will be soil itself, but when the water is to be used for livestock, supplementary irrigation or human consumption, a storage facility of some kind will have to be produced. In countries where land is abundant, water harvesting involves: harvesting or reaping the entire rain water, store it and utilize it for various purposes. In India, it is not possible to use the land area only to harvest water. Hence water harvesting means use the rain water at the place where it falls to the maximum and the excess water is collected and again reused in the same area. Therefore, the meaning of water harvesting is different in different areas/countries. The methods explained here are for both agriculture and to increase the ground water availability by recharging the ground water.

The water harvesting for household and for recharging purposes are also in existence for long years in the world. During rainy days, the people in the villages used to collect the roof water in big vessels and use it for house hold purposes. The people in the rural areas in South East Asian countries used to collect the roof water (Thatched roof by providing gutters) by placing 4 big earthen vessels in 4 corners of their houses. They used this water for all household purposes including drinking and if it is exhausted only they will go for well water. The main building of the Agricultural College at Coimbatore, India was constructed 100 years ago and they have collected all the roof water by pipes and stored in a big under ground masonry storage tanks on both sides of the building. This rain water is used for all labs, which require pure and good quality of water. On the same principle, arrangements were made to collect all the rain water falling on the terrace in all the buildings constructed subsequently. Even the surface water flowing in the Nalla's in the university campus (TNAU) are also diverted by providing obstructions to the abandoned open wells to recharge ground water.

Hence, rain water harvesting is as old as civilization and practiced continuously in different ways for different purposes in the world. However, it has not been done systematically and scientifically in all places. Need has come to harvest the rain water including roof water to solve the water problems everywhere not only in the arid but also in the humid regions.

2.3   NEED FOR RAIN WATER HARVESTING: INDIA

Water is a becoming a scarce commodity and it is considered as a liquid gold in South India. The demand of water is also increasing day by day not only for agriculture, but also for household and Industrial purposes. It is estimated that water needs for drinking and other municipal uses in India will be increased from 3.3 MHm to 7.00 MHm in 2020/25. Similarly the demand of water for industries will be increased by 4 fold, i.e., from 3.0 MHm to 12.00 MHm during this period. At the same time more area should be brought under irrigation to feed the escalating population of the country, which needs more water. But we are not going to get one liter more water than we get at present though the demand is alarming.

The perennial rivers are becoming dry and ground water table is depleting in most of the areas. In Coimbatore – India, the depletion is about 30 m in the last 30–40 years. Country is facing floods and drought in the same year in many states. This is because, no concrete action was taken to conserve, harvest and manage the water efficiently.

The rainfall is abundant in the world and also in India. But it is not evenly distributed in all places. India being the monsoonic country, the rain falls only for 3 to 4 months in a year with high intensity, which results more runoff and soil loss. Total rain occurs only in about 100 hours out of 8760 hours in a year. It is also erratic and it fails once in 3 or 4 years. This is very common in many parts of India.

The availability of water in the world, in India and in Tamil Nadu is shown in Table 2.1.

If the availability of water is 1700 M³/p/y, there will be occasional water stress, and if it is less than 1000 M³/p/y, it is under water scarcity condition as per Dr. Malin Falkenmark, a widely respected Swedish hydrologist. Though India is not under water stress. Yet Tamil Nadu is already under water scarcity condition. However, there is no need for panic since it is possible to manage this condition as in the case of Israel where the availability of water is about 450 M³/p/y, by means of water harvesting, water conservation and water management.

Water scarcity/stress are not limited to arid regions alone, but it is there even in high rainfall areas also. Chirapunji in India gets more than

TABLE 2.1   The Availability of Water

Places	Rainfall, mm	Population	Availability of water, m³/person/year
World	840	7 billion	6500
India	1150	1.25 billion	2000
Tamil Nadu (India)	925	72 million	650

11,000 mm of annual rainfall but it faces drinking water problem before monsoon commences where as in Ralegan Siddhi in Maharashtra state, there is no water scarcity problem though the average annual rainfall is only about 450 mm. Hence to mitigate water problem/drought, etc., there is an urgent need to follow our ancestral way of water harvesting including roof water harvesting and the latest technologies adopted in soil and water conservation measures on watershed basis which are described in this chapter.

In the Theme paper on water vision 2050 of India, Water Resources Society (IWRS) has indicated that a storage capacity of 60 MHm is necessary to meet the demand of water for irrigation, drinking and other purposes. But the present live storage of all reservoirs put-together is equivalent of about 17.5 MHm which is less than 10% of the annual flow in the rivers in India. With the projects under construction (7.5 MHm) and those contemplated (13 MHm) are added, it comes to only 37.5 MHm and hence we have to go a long way to build up storage structures in order to store 60 MHm.

More than 75% of the area comes under hard rock in Tamil Nadu. Further the porosity of the rock is only about 3%. The natural recharge of rain water in this region is only about 10–12%, which is very minimal. Therefore, there is an urgent need to take up the artificial recharge of the rain water for which water harvesting and water conservation structures are to be built up in large scale. The rainfall in coastal area is more than 1200 mm (Chennai), still drinking water is a problem in almost every year. This is because the entire rain water is collected in masonry drains (from houses, streets/roads, etc.) are taken to the sea instead of taking into the ground water aquifers. The ground water available can be used during summer and make the aquifer empty so that the rain water can be put into the aquifers during rainy period.

All these details indicate the need for water harvesting measures in urban and rural areas for use of Agriculture, drinking and other purposes.

2.4   PRESENT STATUS AND TECHNIQUES ON WATER HARVESTING

The (rain) water harvesting system for agriculture in semi-arid parts may have the following components or phases depending upon location specific situations:

- direct rain water conservation;
- in situ moisture conservation;
- water harvesting (run off collection and storage); and
- inducements of more run off on non-cropped area and prevention of losses.

2.4.1 DIRECT RAIN WATER CONSERVATION

The approaches to rain water conservation are through agronomic and engineering measures. This will not only harvest, conserve water but also prevent soil erosion particularly in semi-arid parts.The agronomic measures are:Contour farming, strip cropping, farming terraces, cover crops, off season tillage in light and red soils, deep tillage in hard pan areas, summer fallow, mulching to prevent evaporation, vertical mulch and providing vegetative barrier on contour in land.

The engineering measures adopted differ from place to place with reference to slope, soil type, intensity and total rain, etc., Depending upon the parameters the methods followed are, – the contour trenching, contour stone walls, staggered trenching, constructing temporary and permanent check dams, gully plugging, contour bunding, compartmental bunding, land leveling and the like.

2.4.2 IN SITU MOISTURE CONSERVATION

In most of the semi-arid tropics, the rainfall is erratic and falls within a short time. The moisture may not be available to the crop at the critical stage of its growth. In-situ moisture conservation can help in retaining soil moisture regimes for a longer duration. Here again there are very many location – specific techniques and are to be chosen appropriately. In short, they are forming dead furrows, ridging, graded furrows, broad bed and furrows in black soil, tied ridging, water spreading in agricultural cropped area. For tree crops, micro catchments, saucer basin/semi circular basins/ crescent bunds and catch pits can be introduced.

2.4.3   WATER HARVESTING – RUN OFF COLLECTION AND STORAGE

In dry land agriculture, it is collection of excess run off water in a storage tank and using it for betterment of crop production in the collected or other areas. There are three types of collector tanks: farm ponds, percolation ponds and silt detention tanks. The water collected in the farm pond is directly used for protective irrigation. The water stored in other structures will recharge the ground water. This water can be used for supplemental/ protective irrigation by providing wells.

2.4.4   INDUCEMENTS FOR MORE RUN OFF

Modification of ground surface to induce run off is the main aspects of rain water harvesting in arid lands where rain falls in frequently and its quantity is insufficient for good crop growth. Under such conditions, run off may be increased by treating non cropped catchments and diverting to cropped area or storage reservoir for irrigation. The techniques used are: land alteration, soil compaction, soil deflocculants and additives, spraying asphalt membranes, etc., these methods are mostly followed in USA, Australia where land area is plenty compared to population.

2.4.5   ROOF WATER HARVESTING FOR DRINKING AND OTHER MUNICIPAL NEEDS

Rain water harvesting from the roof of the house for drinking and other needs are not new concepts. It is practiced by the people both in villages and towns for many years. Roof water collection has many advantages in urban and low rainfall areas. It is being practiced in both the developed and developing countries – Japan, Philippines, Germany, Caribbean, Thailand, Denmark, China, Kenya and India.

In India, the roof water collection were made 100 years back at the TNAU and the water collected from the roof are stored in the underground tanks and the water is being used for all the labs which requires good quality of water. Rain water harvesting is widely used in North Eastern states.

It is growing popular especially in the rural areas, where piped water supply is in adequate and expensive due to hilly terrain. In Thailand one of the world's leading rain water harvesting nations has launched to supply clean drinking water to rural areas under the UN water supply and sanitation decades (1981–1990). It was aimed in catching in the rain water in jars made of earth. Rain water harvesting in the North East, where it is the only source for drinking water, is underground water.

Roof water harvesting has a great scope to solve drinking water problem in the cities/towns in the country. The government of India is giving importance and encouragement in the roof water harvesting and water recharging activities through the central ground water board. The Chennai metropolitan water supply and sewage board (metro water) has made water harvesting structures are compulsory for multi structural buildings from 1997. It is estimated about 6000 complex in Chennai have roof water harvesting systems. In Coimbatore also any new construction must have water harvesting structures. More than 1000 houses/premises are said to have provided with water harvesting structures. The same can be adopted in all corporations/municipal towns and panchayats in the country which will go a long way to migrate the water scarcity/problems in the country.

2.4.6   ROLE OF INDUSTRIES IN RAIN WATER HARVESTING

The industry can play a meaningful role in accelerating the rain water harvesting program by:

- Organizing mass awareness campaigns in industrial areas by issuing hand-outs, papers and organize village level meetings through NGO's and panchayats.
- Practicing rain water harvesting and recharging in their own factory premises and houses.
- Cleaning the ooranis/village tanks.
- Use the abandoned open/bore wells for recharging.
- Assisting financial support for construction of check dams/percolation tanks.
- Recharging sewage water after treatment (e.g., Los Angeles, U.S.A.).

2.5   METHODS OF WATER HARVESTING IN RURAL AND URBAN AREAS

There are different/various systems of water harvesting depending upon the source of water supply as classified below:

a. In situ rain water harvesting:
- Bunding and terracing
- Vegetative/Stone contour barriers
- Contour trenching
- Contour stone wells
- Contour farming
- Micro catchments
- Tie Ridging methods

b. Direct surface runoff harvesting:
- Roof water collection
- Dug out ponds/storage tanks
- Tanks
- Khadins
- Urani's/Village tanks
- Temple tanks
- Diversion bunds
- Water spreading

c. Stream flow/runoff harvesting:
- Nalla bunding
- Gully control structures
- Check dams – temporary/permanent

d. Permanent:
- Silt detention tanks
- Percolation ponds
- Farm ponds

e. Sub surface flow harvesting:
- Sub surface dams
- Diaphragm dams

f. Micro catchment's/watershed:
- Inter terrace/inter plot water harvesting
- Conservation bench terracing

g. Runoff inducement by surface treatment:
*   Roaded catchment's
*   Use of cover materials – aluminum foils
*   Plastic sheet, bentonite, rubber, etc.
*   Using chemicals for water proofing, water repellent, etc.

This is a comprehensive list of soil and water conservation measures including water harvesting on watershed basis.

2.6 PLAN OF ACTION FOR RAIN WATER HARVESTING

As stated early, rain water harvesting is as old as civilization and is practiced in many countries including India from time immemorial. But government and people remember this only when water is not available even for drinking purposes. There is no use of spending huge sum of money when we notice the water scarcity even for drinking. These activities/structures should be taken/constructed before the rainy season so that the rain water which goes as runoff outside the sub watershed/city limits can be collected and used directly or by recharging into the ground. Government is undertaking the wasteland, watershed development, drought prone area programs, but not done in a comprehensive/integrated manner saturating the watershed in all water harvesting measures on a watershed basis. Hence there is a need to take up watershed development programs in all the watersheds – mainly water harvesting measures in a scientific and systematic manner.

Similarly corporation like Chennai/Coimbatore in India have passed resolution that in any construction, water harvesting work should be included and executed, but in practice, it is not implemented. The authorities concerned should monitor the program so that the drinking water problem can be solved in all municipalities/corporation without any difficulty if roof water harvesting is implemented in letter and sprit.

To sum up, Table 2.2 indicates types of water harvesting systems that can be implemented in different parts of India.

2.7 CASE STUDIES IN WATER HARVESTING

There are numerous case studies available in water harvesting both in rural and urban sectors. In rural areas, it is soil and water conservation measures

TABLE 2.2   Types of Water Harvesting Systems

Region	Types of water harvesting system	Use
Arid plains	Artificial catchments to capture rainfall (Tankas or khadins in Rajasthan)	Drinking
	Tankas or talabs in Rajasthan to capture surface runoff	Drinking and irrigation
	Embankments/obstructions across drainage/Nalla to capture surface runoff	Irrigation water
Semi-arid region	Tanks/Ponds/Eri to capture surface runoff and also chains of tanks called cascade	Irrigation water and drinking water through recharge of ground water
Flood plains	Mud embankment which may be breached during the floods	Irrigation water and drinking water through recharge of ground water
Hill and mountain region	Diverted stream flows Jammu, M.P., Maharashtra	Irrigation water

taken on watershed basis to conserve soil moisture and augment ground water. In the urban sectors, it is mostly roof water harvesting for direct use and recharging the ground water and also collecting of surface runoff from pavements/roads and recharging it into the ground through recharge pits or using abandoned/existing wells. The following are the places where rain water/roof water harvesting have been implemented in a successful manner in India and abroad.

   i.   Rural areas: Soil and water conservation and water harvesting measures:
   •   Lakshman Nagar/Varisai Nadu in Theni District, Tamil Nadu.
   •   Ralegan Siddhi in Maharashtra state.
   •   Alankulam Taluk in Tirunelveli District, Tamil Nadu.
   •   Aravari watershed in Alwar District, Rajasthan.
   •   Maheshwaram watershed in Andhra Pradesh.
   •   Kapilnalla watershed and Rajanukunte – Mittemari micro watershed in Karnataka State.
   •   Jam Nagar District and Kutch District in Gujarat in Different places.
   •   Sukhomajri near Chandigarh, Punjab.

- Waste land/fallow land, dry land-watershed development works in TN, AP, MP, etc.

ii. Urban sectors: Mostly the roof water harvesting for direct use and recharging:
  - Tamil Nadu Agricultural University, Coimbatore, all main buildings.
  - Pricol Industry In Periyanaickenpalayam, Coimbatore District (Industry building).
  - TWAD Board and PWD offices at Chennai.
  - Numerous Apartment buildings in Chennai.
  - Sundaram and Clayton Ltd, Padi, Chennai (Industry buildings).
  - TVS Training school at Vanagaran, Chennai.
  - Rastrapathi Bhavan, New Delhi.
  - Center for science and environment buildings at Delhi.
  - Institute of economic growth, New Delhi.
  - Number of residences at VasanthVikhar, New Delhi.
  - Aizawal, Mizoram.

iii. Foreign countries (List may not be complete):
  - Thailand – Thatched houses in many villages.
  - Tokyo's Sumo wrestling area, Japan.
  - Public buildings in Germany.
  - Singapore – Harnessing of surface runoff and roof water in Changai airport and many high raise buildings.
  - Different types of water harvesting methods in rural and urban areas in Sri Lanka.

Rules and regulations have been framed for rain water harvesting in Chennai and Coimbatore corporations in Tamil Nadu. The Gujarat government has issued a general resolution for the effort that no new construction would be allowed if it does not have provision for roof top rain water harvesting. This would be valid in all 143 municipalities and 6 urban development authorities in the state. It is heartening to note that Confederation of Indian Industries (CII) and Federation of Indian Chambers of Commerce and Industries (FICCI) have taken action to implement the rain water harvesting in premises of their Industry.

Soil and water conservation measures and watershed development works including various special area programs are being implemented in all states. The main objective of these programs are conserving, harvesting and managing the rain water to get more moisture for growing crops and recharging the ground water.

If the above measures are implemented in rural and urban areas, the drought in rural areas and drinking water problem in urban and rural population can be solved. The people, NGO and Government should join together and implement the Rain water harvesting in a big way in all places in the years to come to solve water scarcity problem in the country.

2.8   WATER HARVESTING, CONSERVATION AND MANAGEMENT: INDIA

India is blessed with abundant water resources. The average annual rainfall of the world is about 840 mm whereas in India, it is 1150 mm and in Tamil Nadu it is 925 mm. The availability of water in India per person is about 2000 m^3. This is more than the standard fixed by UN agencies and World Bank which is about 1700 m^3. If the availability is less than 1000 m^3, it is considered as a water scarcity state/nation. In fact, Tamil Nadu is a water scarcity state since the availability of water is about 650 m^3 per person per year. The problem of water scarcity/drought is man-made as we are not harvesting, conserving and managing our available water resources scientifically and taking into account India as one country.

The Government of India has stated that 12 states are under drought now. This is after 12 years of good monsoon in the country. Water scarcity or drought cannot be there if we have properly/scientifically managed the rain water/irrigation Water. Water scarcity should occur only if continuous deficit rainfall is there for 2–3 years. In Ralegan Siddhi, water harvesting and management are done systematically under the dynamic leadership of Mr. Anna Hazare and hence no scarcity problem though the rainfall is very meager (400/450) where as in Cherrapunji where the rainfall is 11,000 mm, there is no such measures followed and hence the problem of drinking water. Therefore, water problem is not limited to the arid climate regions only, but in area with good rainfall, the access to safe water is becoming critical as stated above. The reasons of water scarcity are:

- lack of storage facilities;
- erratic rainfall;
- low infiltration rate; and
- pumping ground water without any consideration.

In Monsoonic countries like India, we must expect all this. Rainfall will not be uniform in all years like Europe. Further there may be failure of rains once in 3 or 4 years and in some years the rainfall is much more than the average rainfall. Because of this phenomenon, the average rainfall is calculated taking the total rainfall for 30 years (a cycle) and average is worked out. Further the rainfall is only there for 3/4 months (monsoon period) and all the rain occurs in about 100 hours in this period. Therefore, we have to conserve/harvest the water to prevent water scarcity/drought to mitigate problem.

Our forefathers constructed numerous tanks, ooranis (village tanks) and temple tanks, farm ponds to store the rain water for irrigation and community uses and recharging into the ground. We are not in a position to maintain these structures. We cry about water only when there is no water even for drinking and we don't bother to conserve water during good rainy years and we leave all the rain water to go as run off to the sea. Though the total flow in all the rivers in India is about 187 MHM (Million Hectare Meter = MHM), we are utilizing only about 40–50 MHM of water and the rest are wasted even after 65 years of Independence.

In Tamil Nadu, the 39,000 tanks were irrigating about one million Ha of land in 1960's and 1970's but due to mismanagement and not maintaining the tank, the capacity has reduced and the tanks are irrigating only about 0.6 million Ha now. In most of the tanks, the tank bed (water spread area) is silted up heavily and the capacity of the tank is reduced, and also they are not maintained properly.

Therefore we have to take action to conserve, harvest and manage the rain water systematically and scientifically. In India the total storage capacity in big, medium, small dams and tanks is only about 18 MHM and the works under progress can store about another 7.5 or 8 MHM which is very less compared to the total quantity of about 187 MHM. In USA, the storage capacity is much more than 60 MHM where the available water is equal to that of India, but having a population of about ¼ of India.

There are three types of water harvesting measures in the urban sectors: Roof water harvesting; Storm water harvesting; and Waste water (sewage) harvesting. The roof water harvesting as detailed in this chapter has just started in Chennai and Coimbatore to some extent, but not in any other cities/towns of India. Similarly storm water coming down from paved areas, tar/concrete roads should be collected in low grounds and it should be recharged. We are interested in drinking water schemes, but we don't bother about the waste/sewage water collection and utilization. It is high time to take action to collect and reclaim the waste/sewage water which can be used for agriculture/irrigation as in the case of Israel where 35% irrigation water is only from sewage/waste water and in 2015/2020 all the irrigation water will be only from sewage/waste water. The waste/sewage water from Coimbatore city alone can irrigate about 20,000 acres if it is properly collected reclaimed and reused.

In rural areas, roof water harvesting can be taken up. But the important activities should be water shed development works in order to conserve, harvest and manage the rain water. Numerous examples are there in India. Here, the main idea is to arrest the run off and store it by constructing check dams, farm ponds percolation ponds and through soil and water conservation measures on water shed basis including in situ moisture conservation measures like forming basins, ridges and furrows, water spreading, catch pits, etc.

In the water management side, to use the ground water efficiently and to maintain ground water level at reasonable depth, water management practices including sprinkler and drip irrigation methods should be introduced in large scale to save water and to increase the productivity and income the farmers. Cultivation of high water consumable crops like paddy, sugarcane, and banana should be banned under well irrigation and drip/sprinkler irrigation only should be adopted if these crops are cultivated. This will go a long way in increasing the production and income at the same time, maintain the ground water table without depletion.

If the suggestions by the author in this chapter are followed, the water problems in India can be solved. For this, the co-operation of the farmers, public, NGO's and Government should be ensured. Appendix A shows illustrations indicating techniques for rain water harvesting (Figures A1–A8).

2.9   SUMMARY

It is very important to make water everybody's business. It means a role for everybody with respect to water. Every household and community has to become involved in the provision of water and in the protection of water resources. Make water the subject of a people's movement. It means the empowerment of our urban and rural community, i.e., to manage their own affairs with the state playing a critical supportive role.

Further, involving people will give the people greater ownership over the water projects including watershed development, soil and water conservation. Water harvesting will go a long way towards reducing misuse of government funds. It will also develop a sense of ownership of water supply system/project so that they will also take good care of water. In this way it is possible to solve water problems facing the country in the 21st century.

KEYWORDS

- catchments
- creek flow
- evaporation
- farm water
- flood diversion
- fresh water
- hydrophobic
- infiltration
- irrigation use
- precipitation
- rain water
- rain water harvesting
- reservoir
- runoff

- safe water
- soil compaction
- storage reservoir
- University of Sydney
- USDA
- warping
- water
- water harvesting
- water scarcity
- water storage capacity
- watershed

REFERENCES

1. Ake, Nilsson (1988). Ground water dams for small-scale water supply. IT publication.
2. Boers, T. M., & Ben-Asher, J. (1982). A review of rainwater harvesting. Agriculture Water Management, 5, 145–158.
3. Center for Science and Environment (2001). Making water everybody's business. New Delhi.
4. Center for Science and Environment (2001). A Water Harvesting Manual. Delhi.
5. Chitale, M. A. (2000). A blue revolution, Bhavans Book University, Pune.
6. CII (2000). Rain water harvesting – A guide. New Delhi.
7. Gandhi, Rajiv (1998). National Drinking water missions Handbook on Rain Water Harvesting. Government of India, New Delhi.
8. Gould, J. E. (1992). Rainwater catchment systems for household water supply. In: Environmental Sanitation Reviews No. 32, ENSIC, Asian Institute of Technology, Bangkok.
9. Hatibu, N., & Mahoo, H. (1999). Rainwater harvesting technologies for agricultural production: A case for Dodoma, Tanzania. In: Conservation Tillage with Animal Traction. Kaumbutho, P. G., & Simalenga, T. E. (eds.). A resource book of the Animal Traction Network for Eastern and Southern Africa (ATNESA). Harare. Zimbabwe. 173 pp.
10. http://www.climatetechwiki.org/content/rainwater-harvesting.
11. http://www.rainwaterharvesting.org/whatiswh.htm.
12. https://en.wikipedia.org/wiki/Rainwater_harvesting.

13. Joshi, P. K., Jha, A. K., Wani, S. P., Joshi, L., & Shiyani, R. L. (2005). Meta analysis to assess impact of watershed program and people's action. Comprehensive Assessment Research Report 8, International Water Management Institute, Colombo.

14. Keith, R. (1975). Water Harvesting: State of art. Watershed management symposium, ASCE irrigation and drainage division, Logan, USA.

15. Samra, J. S. (1996). Water harvesting and recycling – Indian experiences. Central Soil and Water Conservation Research and Training Institute, Dehradun (UP).

16. Sivanappan, R. K. (1999). Soil and water conservation and water harvesting. Tamil Nadu Afforestation Project, Chennai.

17. Sivanappan, R. K. (2001). Water Harvesting. ICCI, Coimbatore.

18. Stockholm Water Symposium (1998). Water harvesting. Aug.

19. UNEP (1982). Rain and storm water harvesting in rural areas, Tycooly International Publishing Ltd., Dublin.

20. UNEP (1997). Sourcebook of alternative technologies for freshwater augmentation in some countries in Asia. UNEP, Unit of Sustainable Development and Environment General Secretariat, Organization of American States, Washington, D.C.

21. UNEP and SEI (Stockholm Environment Institute) (2009). Rainwater harvesting: a lifeline for human well-being, United Nations Environment Program and Stockholm Environment Institute.

22. UNFCCC (2008). National Adaptation Programs of Action. Summary of Projects on Water Resources identified in submitted NAPAs as of September 2008, United Nations.

23. Verma, H. N., & Tiwan, K. N. (1995). Current Status and Prospects of Rain Water Harvesting. NIH, Roorkee, India.

24. Water Aid Technical Brief: Rainwater Harvesting.

APPENDIX A
ILLUSTRATIONS SHOWING TECHNIQUES FOR RAIN WATER HARVESTING

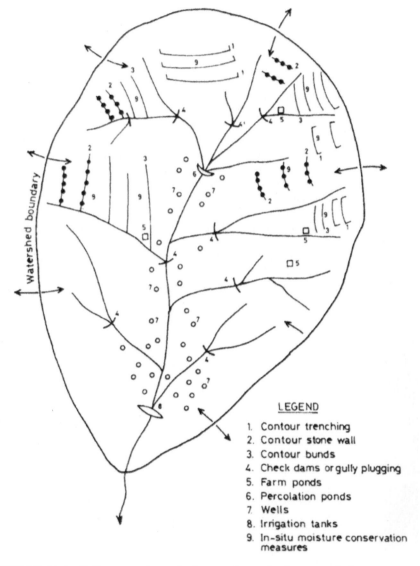

LEGEND

1. Contour trenching
2. Contour stone wall
3. Contour bunds
4. Check dams or gully plugging
5. Farm ponds
6. Percolation ponds
7. Wells
8. Irrigation tanks
9. In-situ moisture conservation measures

FIGURE A2.1   Soil conservation and rain water harvesting in a watershed.

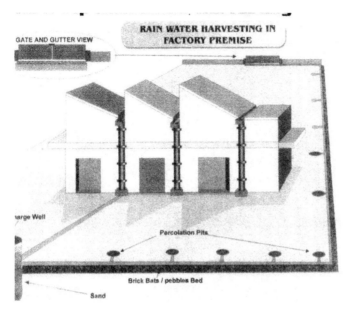

FIGURE A2.2   Rain water harvesting for the roof top in a factory premise.

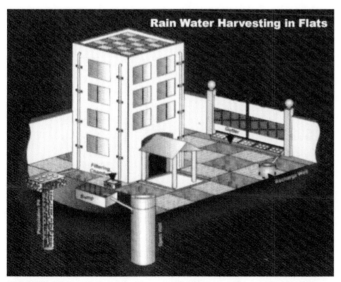

FIGURE A2.3   Rain water harvesting for a multi – storied building or flats.

Elements of a typical water harvesting system

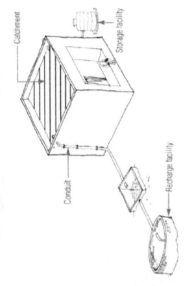

Catchment

Storage facility

Conduit

Recharge facility

Elements of typical water harvesting system

1. Catchments

The catchment of a water harvesting system is the surface which receives rainfall directly and contributes the water to the system. It can be a paved area like a terrace or courtyard of a building, or an unpaved area like a lawn or open ground. Temporary structures like sloping sheds can also act as catchments.

2. Conduits

Conduits are the pipelines or drains that carry rainwater from the catchment or rooftop to the harvesting system. Conduits may be of any material like polyvinylchloride (PVC), asbestos or galvanized iron (GI) materials that are commonly available.

Broadly, rainwater can be harvested for two purposes:

RAINWATER

Stored for ready use in containers above ground or below ground

Charged into soil for withdrawal later (groundwater recharging)

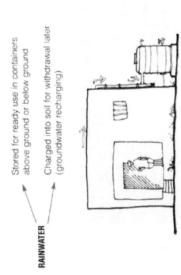

Rainwater can be stored in tanks

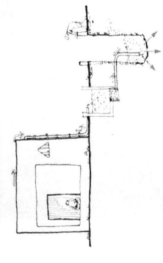

Rainwater can be recharged into the ground

FIGURE A2.4   Process of rain water harvesting.

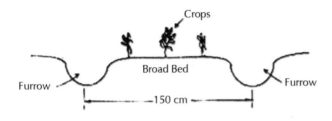

a) BROAD BED AND FURROWS FOR AGAL CROPS

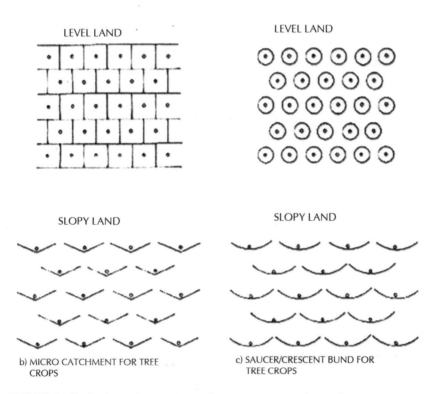

b) MICRO CATCHMENT FOR TREE CROPS

c) SAUCER/CRESCENT BUND FOR TREE CROPS

FIGURE A2.5 In situ moisture conservation measures: water harvesting.

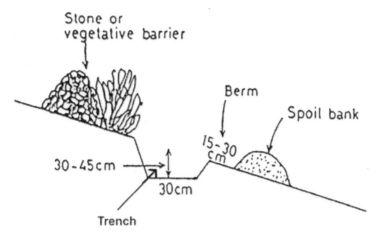

FIGURE A2.6   Cross section of a contour trench.

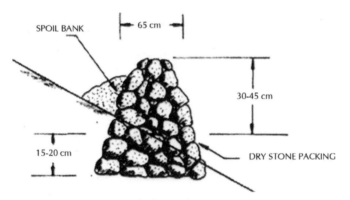

FIGURE A2.7   Contour stone wall.

FIGURE A2.8   Temporary check dam: water harvesting.

CHAPTER 3

PERCOLATION TANKS FOR WATER CONSERVATION FOR SUSTAINABLE DEVELOPMENT OF GROUND WATER IN HARD-ROCK AQUIFERS

SHRIKANT DAJI LIMAYE

CONTENTS

3.1 INTRODUCTION

Development of a natural resource like ground water is a concerted activity towards its sustainable use for human benefit. The concept of sustainable

Modified and printed from: *Shrikant Daji Limaye (2011). Importance of Percolation Tanks for Water Conservation for Sustainable Development of Ground Water in Hard-Rock Aquifers in India. Chapter 2, In: Water Conservation, Dr. Manoj Jha (Ed.), ISBN: 978-953-307-960-8, InTech, Available from: www.intechopen.com/books/water-conservation/importance-of-percolation-tanks-for-water-conservation-for-sustainable-development-of-ground-water-i.*

use is related to various factors like the volume of water storage in the aquifer, annual recharge or replenishment, volume of annual pumpage for the proposed use, benefit-cost ratio of the proposed use, and environmental impacts of the proposed use.

In this chapter, hard rock aquifers mean the non-carbonate, fractured rocks like the crystalline basement complex and metamorphic rocks, which cover an area of about 800,000 sq. km. in central and southern India. Basalts of western India also known as the Deccan traps of late Cretaceous to early Eocene period are also included as a special case. Deccan traps comprise hundreds of nearly horizontal, basaltic lava flows in a thick pile and cover around 500,000 sq. km. of western India (Figures 3.1a and 3.1b). This pile was not tectonically disturbed after consolidation and a hand specimen does not show any primary porosity due to the non-frothy nature of the lava [1]. Hydrogeologically, the Deccan traps have low porosity and are therefore, akin to fractured hard rock aquifers. The most significant features of the hard rock aquifers are as follows:

1. A topographical basin or a sub-basin generally coincides with ground water basin. Thus, the flow of ground water across a prominent surface water divide is very rarely observed. In a basin, the ground water resources tend to concentrate towards the central portion, closer to the main stream and its tributaries.

2. The depth of ground water occurrence, in useful quantities, is usually limited to a hundred meters or so.

3. The aquifer parameters like storativity (S) and transmissivity (T) often show erratic variations within small distances. The annual fluctuation in the value of T is considerable due to the change in saturated thickness of the aquifer from wet season to dry season. When different formulae are applied to pump-test data from one well, a wide range of values of S and T are obtained. The applicability of mathematical modeling is limited to only a few simpler cases within a watershed. But such cases do not represent conditions over the whole watershed.

4. The phreatic aquifer comprising the saturated portion of the mantle of weathered rock or alluvium or laterite, overlying the hard fractured rock, often makes a significant contribution to the yield obtained from a dug well or bore well.

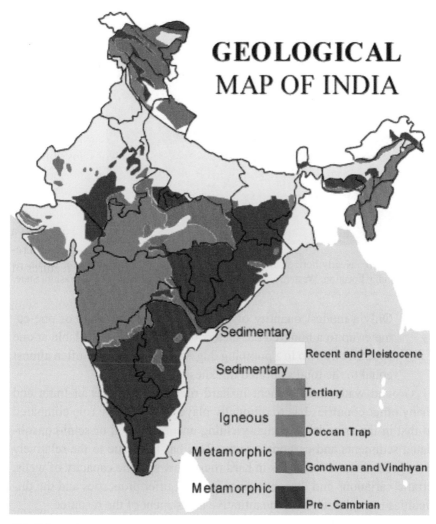

FIGURE 3.1A Geological map of India showing basement complex and basalts of peninsular region, (Scale: 1 cm = 200 km). Note: Basement complex (Granite, Gneiss, and other Metamorphic rocks) is shown in brown color. Basaltic area is shown in green. Black lines show State boundaries. The state of Maharashtra on the western coast is almost covered by basalt or Deccan trap (Green color). Peninsular India is mostly covered by basalt and basement complex.

FIGURE 3.1B   Nearly horizontal lava flows comprising the Deccan traps or basalts of Western India. Location: Western Ghat hills between Pune and Mumbai, Maharashtra State.

5. Only a modest quantity of ground water, in the range of one cu. meter up to a hundred cu. meters or so per day, is available at one spot. Drawdown in a pumping dug well or bore well is often almost equal to the total saturated thickness of the aquifer.

Ground water development in hard rock aquifer areas in India and many other countries has traditionally played a secondary role compared to that in the areas having high-yielding unconsolidated or semi-consolidated sediments and carbonate rocks. This has been due to the relatively poor ground water resources in hard rocks, low specific capacity of wells, erratic variations and discontinuities in the aquifer properties and the difficulties in exploration and quantitative assessment of the resource.

It should, however, be realized that millions of farmers in developing countries have their small farms in fractured basement or basaltic terrain. Whatever small supply available from these poor aquifers is the only hope for these farmers for upgrading their standard of living by growing irrigated crops or by protecting their rain-fed crops from the vagaries of Monsoon rainfall. It is also their only source for drinking water for the family and cattle. In many developing countries, like in India, hard rock hydrogeologists have, therefore, an important role to play.

3.2   OCCURRENCE OF GROUND WATER

Ground water under phreatic condition occurs in the soft mantle of weathered rock, alluvium and laterite overlying the hard rock. Under this soft mantle, ground water is mostly in semi-confined state in the fissures, fractures, cracks, and joints [2]. In basaltic terrain, the lava flow junctions and red boles sandwiched between two layers of lava flows, also provide additional porosity (Figure 3.2). The ratio of the volume of water stored under semi-confined condition within the body of the hard rock, to the volume of water in the overlying phreatic aquifer depends on local conditions in the mini-watershed. Dug-cum-bored wells tap water from the phreatic aquifer and also from the network of fissures, joints and fractures in the underlying hard rock (Figures 3.3a and 3.3b).

FIGURE 3.2   Red bole (intertrappean bed) sandwiched between hard, fractured basalt flows (Exposure seen in road-side cutting on Pune-Bangalore highway Pune District, Maharashtra State).

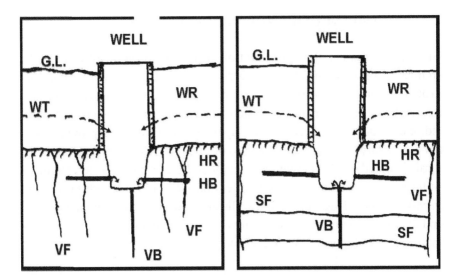

FIGURES 3.3 3a(Left) and 3.3b(Right). Dug cum bored wells. Legend: GL – Ground level, HB – Horizontal bore, HR – Hard rock, SD – Sheet fracture or joint, VB – Vertical bore, VF – Vertical fracture, WR – Weathered rock, WT – Water table.

3.2.1 THE RECHARGE TO GROUND WATER

The recharge to ground water takes place during the rainy season through direct infiltration into the soft mantle overlying the hard rock and also into the exposed portions of the network of fissures and fractures. In India and other Asian countries in Monsoon climate, the ratio of recharge to rainfall in hard rock terrain is assumed between 3 and 15% [5]. This ratio depends upon the amount and nature of precipitation, the nature and thickness of topsoil and weathered zone, type of vegetation, evaporation from surface of wet soil, profile of underlying hard rock, the topographical features of the sub-basin and the status of soil and water conservation activities adopted by villagers. Ground water flow rarely occurs across the topographical water divides and each basin or sub-basin can be treated as a separate hydrogeological unit for planning the development of ground water resources. After the rainy season, the fully recharged hard rock aquifer gradually loses its storage mainly due to pumpage and effluent drainage by streams and rivers. The dry season flow of the streams is thus supported by ground water outflow. The flow of ground water is from the peripheral portions of a sub-basin to the central-valley portion, thereby causing de-watering of the portions closer to topographical water divides. In many cases, the dug wells and bore wells yielding perennial supply of ground water can only be located in the central valley portion.

The annual recharge during Monsoons being a sizable part of the total storage of the aquifer, the whole system in a sub-basin or mini-basin, is very sensitive to the availability of this recharge. A couple of drought years in succession could pose a serious problem. The low permeability of hard rock aquifer is a redeeming feature under such conditions because it makes small quantities of water available, at least for drinking purpose, in the dug wells or bore wells in the central portion of a sub-basin. If the hard rocks had very high permeability, the ground water body would have quickly moved towards the main river basin, thereby leaving the tributary sub-basins high and dry. The low permeability in the range of 0.05 to 1.0 meter per day thus helps in retarding the outflow and regulating the availability of water in individual farm wells. More farmers are thus able to dig or drill their wells and irrigate small plots of land without causing harmful mutual interference.

3.3 GROUND WATER DEVELOPMENT

In the highly populated but economically backward areas in hard rock terrain, Governments in many developing countries have taken up schemes to encourage small farmers to dig or drill wells for small-scale irrigation. This is especially true for the semi-arid regions where surface water resources are meager. For example, in peninsular India, hard rocks such as granite, gneiss, schist, quartzite (800,000 sq. km.) and basalts (Deccan traps- 500,000 sq. km.) occupy about 1.30 million sq. km. area out of which about 40% is in semi-arid zone, receiving less than 750 mm rainfall per year. Over 4.00 million dug wells and bore wells are being used in the semi-arid region for irrigating small farm plots and for providing domestic water supply.

Development of ground water resources for irrigational and domestic use is thus a key factor in the economic thrift of vast stretches of semi-arid, hard rock areas. The basic need of millions of farmers in such areas is to obtain an assured supply for protective irrigation of at least one rain-fed crop per year and to have a protected, perennial drinking water supply within a reasonable walking distance. The hard-rock hydrogeologists in many developing countries have to meet this challenge to impart social and economic stability to the rural population, which otherwise migrates to the neighboring cities. The problem of rapid urbanization by exodus of rural population towards the cities, which is common for many developing countries, can only be solved by providing assurance of at least one crop and rural employment on farms.

Ground water development in a sub-basin results in increased pumpage and lowering of the water table due to the new wells, resulting in the reduction of the effluent drainage from the sub-basin. Such development in several sub-basins draining into the main river of the region reduces the surface flow and the underflow of the river, thereby affecting the function of the surface water schemes depending on the river flow. In order to minimize such interference, it is advisable to augment ground water recharge by adopting artificial recharge techniques during rainy season and also during dry season. The measures for artificial recharge during Monsoon rains include contour trenching on hill-slopes, contour bunding of farms, gully plugging, farm-ponds, underground stream bunds, and forestation of barren lands with suitable varieties of grass, bushes and trees. Artificial recharge in dry season is achieved through construction of percolation tanks.

However, increase in pumpage takes place through the initiative of individual farmers to improve their living standard through irrigation of high value crops, while recharge augmentation is traditionally considered as Government's responsibility and always lags far behind the increase in pumpage. In many parts of the world, particularly in developing countries, groundwater is thus being massively over-abstracted. This is resulting in falling water levels and declining well yields; land subsidence; intrusion of salt water into freshwater supplies; and ecological damages, such as, drying out wetlands.

Groundwater governance through regulations has been attempted without much success, because the farmers have a strong sense of ownership of ground water occurring in their farms. Integrated Water Resources Development (IWRM) is being promoted as a policy or a principle at national and international levels but in practice at field level, it cannot be attained without cooperation of rural community. NGOs sometimes play an important role in educating the villagers and ensure their cooperation.

3.4 IMPORTANCE OF DRY SEASON RECHARGE

During the rainy season from June to September, the recharge from rainfall causes recuperation of water table in a sub-basin from its minimum level in early June to its maximum level in late September. This is represented by the equation:

$$P = R + ET + r \qquad (1)$$

where, P is the precipitation, R is surface runoff, ET is evapotranspiration during the rainy season and r is the net recharge = the difference between minimum storage and maximum storage in the aquifer.

However, after the aquifer gets fully saturated, the additional infiltration during the Monsoons is rejected and appears as delayed runoff. During the dry season, depletion of the aquifer storage in a sub-basin, from its maximum value to minimum value, is represented by the following equation:

[Aquifer storage at the end of rainy season, i.e., maximum storage] = [Aquifer storage at the end of summer season, i.e., minimum storage] + [Pumpage, mainly for irrigation, during the dry season from dug wells and bore wells] + [Dry season stream flow and underflow supported by ground water] − [Recharge, if any, available during the dry season, including the return flow from irrigated crops] (2)

The left-hand side of the above equation has an upper limit, as mentioned above. On the right-hand side, the minimum storage cannot be depleted beyond a certain limit, due to requirement for drinking water for people and cattle. Dry season stream flow and underflow supported by ground water have to be protected, as explained earlier, so that the projects depending upon the surface flow of the main river are not adversely affected. Any increase in the pumpage for irrigation during dry season due to new wells must therefore be balanced by increasing the dry season recharge.

The best way to provide dry season recharge is to create small storages at various places in the basin by bunding gullies and streams for storing runoff during the rainy season and allowing it to percolate gradually during the first few months of the dry season. Such storages created behind earthen bunds put across small streams are popularly known as percolation tanks (Figures 3.4 and 3.5). In semi-arid regions, an ideal percolation tank with a catchment area of 10 sq. km. or so, holds maximum quantity by end of September and allows it to percolate for next 4 to 5 months of winter season. Excess of runoff water received in Monsoon flows over the masonry waste weir constructed at one end of the earthen bund. By February or March the tank is dry, so that the shallow water body is not exposed to high rates of evaporation in summer months (Figure 3.6). Ground water movement being very slow, whatever quantity percolates between Octo-

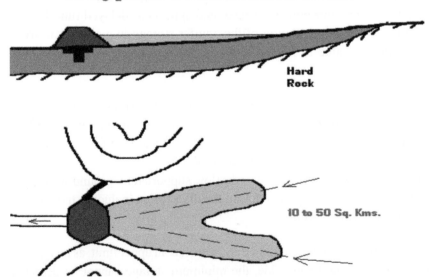

A Typical Percolation Tank

Hard
Rock

10 to 50 Sq. Kms.

FIGURE 3.4   Cross section and plan of a typical percolation tank on a stream between two hillocks. Hillocks are represented by black contours. Earthen bund of the tank is shown in brown; Cut-off trench below the bund is in black and accumulated rain water is in pale blue color.

FIGURE 3.5 Stone Pitching on the face of the earthen bund of a percolation tank under construction. Photo from village Hivre Bazar, District: Nagar, Maharashtra state.

FIGURE 3.6 A percolation tank about to get dry towards beginning of summer. Location: Village: Ralegan Siddhi, District: Nagar, Maharashtra State.

ber and March, is available in the wells on the downstream side of the tank even in summer months till June or the beginning of next Monsoon season. Irrigation of small plots by farmers creates greenery in otherwise barren landscape of the watershed (Figure 3.7).

Studies carried out in granite-gneiss terrain have indicated that about 30% of the stored water in the tank percolates as recharge to ground water in the dry season. The efficiency is thus 30%. In basaltic terrain, if the tank is located at suitable site and the cut-off trench in the foundation of tank-bund does not reach up to the hard rock, higher efficiencies up to 70% could be obtained [4]. However, more research is required for estimation of the impact of percolation tanks in recharge augmentation. In the state of Maharashtra in western India, over 10,000 percolation tanks have been constructed so far [3]. They are beneficial to the farmers and are very popular with them.

The initial efficiency of a percolation tank reduces due to silting of its bottom by receiving muddy runoff from the watershed. If the watershed is well-forested and has a cover of grass, bushes and crops, the silting is minimal. But in an average of 5 to 6 Monsoon seasons the tank bed accumulates about 0.20 to 1.00 meters of silt. Silt reduces

FIGURE 3.7   Greenery crated within a dry, semi-arid watershed with the help of water conservation, farm ponds and percolation tank. The open well has shallow water table even in summer. A low cost centrifugal pump would soon be installed for small scale irrigation. (Location: Village Hivre Bazar. Dist: Nagar, Maharashtra State).

the storage capacity of the tank and also impedes the rate of vertical flow of recharge because of its low permeability. The efficiency gets reduced due to silting and de-silting of tank bed when it dries in summer, becomes necessary [6].

Another type of recharge available during the dry season is the return flow or the percolation below the root zone of crops, from irrigated farms. This return flow to ground water is usually estimated at about 25% to 30% of the volume of ground water pumped in dry season and applied for irrigation. However, due to increasing popularity of more efficient irrigation methods like sprinkler or drip systems, this type of recharge has a declining trend.

3.5 SUMMARY

A watershed is the meeting point of climatology and hydrology. It is therefore, necessary to manage our watersheds so as to absorb the climatic shocks likely to come from the erratic climatic patterns expected in near future. This can be done only through practicing soil and water conservation techniques for artificial recharge during rainy season and through construction of small percolation tanks for artificial recharge during the dry season.

Basin or Sub-Basin management begins with soil and water conservation activities taken up with people's active participation in several sub-basins within a large basin. This improves the shape of hydrograph of the stream or river in the basin, from a 'small time-based and sharp-peaked hydrograph' to a 'broad time-based and low-peak hydrograph.' Such a change also increases ground water recharge.

Small water storages or tanks created in the sub-basins by bunding streams and gullies, store runoff water during the Monsoon season and cause recharge to ground water during the next few months of dry season. The residence time of water in the basins is thus increased from a few months to a few years and the percolated water is available in the wells even during the summer season of a drought year.

After a few years of operation, silting of the tank bed reduces the volume of water stored and also the rate of vertical infiltration. Regular desilting of tanks by local people is, therefore, advisable.

A national policy for afforestation of degraded basins with proper species of grass, bushes and trees should be formulated. Afforestation with eucalyptus trees should not be encouraged in low rainfall areas as this effectively reduces ground water recharge. The main aim of forestation of a degraded watershed with local spices of hardy trees, grasses, etc. should be to conserve soil, reduce velocity of runoff water, promote recharge to ground water and increase the biomass output of the watershed.

Involvement of NGOs should be encouraged in forestation schemes and soil & water conservation programs so as to ensure active participation of rural community in recharge augmentation. NGOs also motivate the farmers to maintain the soil and water conservation structures put in by Government Departments so as to ensure long-term augmentation of recharge to ground water. Along with such management on supply side, demand management is also equally important. NGOs play a significant role in promoting the use of efficient irrigation methods and selection of crops with low water requirement.

The website www.igcp-grownet.org of the UNESCO-IUGS-IGCP Project GROWNET (Ground Water Network for Best Practices in Ground Water Management in Low-Income countries) gives several best practices including soil and water conservation, recharge augmentation, etc. for sustainable development of ground water. The author of this chapter is a Project Leader of GROWNET [7]. The reader is advised to visit the website for detailed information.

Although the discussion in the chapter refers to hard rock terrain in India, it would be equally applicable to many other developing countries, having a similar hydro-geological and climatic set-up.

KEYWORDS

- aquifer
- aquifer storage
- basaltic lava
- basalts
- basement complex

- **benefit-cost ratio**
- **bore well**
- **Deccan traps**
- **demand management**
- **drawdown**
- **dug well**
- **evapotranspiration**
- **ground water**
- **ground water management**
- **GROWNET**
- **hard-rock aquifer**
- **infiltration**
- **net recharge**
- **NGOs**
- **percolation tank**
- **phreatic aquifer**
- **porosity**
- **precipitation**
- **runoff**
- **specific capacity**
- **storativity**
- **sustainable development**
- **transmissivity**
- **water storage**
- **weathered rock**

REFERENCES

1. Adyalkar, P. G., & Mani, V. V. S. (1971). Paloegeography, geomorphological setting and groundwater possibilities in Deccan Traps of western Maharashtra. Bulletin Volcanol., 35, 696–708.
2. Deolankar, S. B. (1980) The Deccan basalts of Maharashtra State, India: Their potential as aquifers. Ground Water, 18(5), 434–437.

3. DIRD website: Efficiency of percolation tanks in Jeur sub Basin of Ahemadnagar District, MS, India. (Visited June 2011) www.dird-pune.gov.in/rp_PercolationTank.htm
4. Limaye, D. G., & Limaye, S. D. (1986). Lakes & Reservoirs: Research and Management, 6, 269–271.
5. Limaye, S. D., & Limaye, D. G. (1986). Proc. of International Conference on ground water systems under stress. Australian Water Resources Conference Series. 13, 277–282.
6. Limaye, S. D. (2010). Review: Groundwater development and management in the Deccan Traps (Basalts) of western India. Hydrogeology Journal, 18, 543–558.
7. http://www.igcp-grownet.org.

CHAPTER 4

PRECISION FERTILIZER APPLICATOR: LOCALIZED AND SUBSURFACE PLACEMENT

THANESWER PATEL and G. R. RAMAKRISHNA MURTHY

CONTENTS

4.1   INTRODUCTION

Agriculture has been a major contributor to national economy of India, especially during the last four decades with the advent of green revolution that ushered in the giant leap of progress in the area of crop production. The agriculture sector today contributes nearly 35% of net national product. During the past few years, agriculture sector has witnessed spectacular advances in the production and productivity of food grains. Rice is one of the major food grains of India and mostly grown in kharif season. Moreover, demand of food grain is increasing day-by-day due to increasing population.

It is estimated that Indian population will reach 1400 million by 2025, requiring 300 million tons of food grain. In order to achieve the required food grains need of the population, crop productivity and production will need to increase and improve through appropriate planning and best utilization of chemical fertilizers and seeds. About 70% of growth in the food production can be attributed through proper planning and optimum utilization of resources these resources. Fertilizer is one of the essential components of modern agriculture. India is the second largest consumer of fertilizers in the World, after China. It accounted for 15.3% of the World's consumption of nitrogenous (N), 19% of phosphatic (P) and 14.4% of potassic (K) nutrients [1]. Due to steady increase in cost of chemical fertilizers owing to fast depletion of petroleum all over the world, the traditional practice of broadcasting the fertilizer which involves a lot of wastage of fertilizer is no more feasible.

There is a need to improve the utilization of fertilizer application in rice with a view to reduce the cost of production and environmental threat due to emission of methane associated with it. Placement of fertilizer near the seed at the time of planting rather than uniform distribution in entire area by various methods of application such as deep placement [2] and banding [3] improves the yield per unit quantity of fertilizer applied and also reduces the wastage of fertilizer. Relative recoveries and levels of N loss can also be influenced by fertilizer composition, and the rate or timing of application [4–6].

Decreasing Ammonia (NH_3) loss by alternative methods of application to surface broadcasting (i.e., incorporation, burying at depth) is related to the provision of a physical barrier in the form of a layer of soil to trap the NH_3 liberated, and the influence of these methods on the ammonical – N concentrations in surface soil solution or floodwater. To ensure effective utilization of nitrogen, deep placement is advocated to reduce the losses due to surface runoff, leaching and volatilization. Mechanical placement of fertilizer [7] through a suitable device is only alternative to achieve this goal. The various farm machinery used for application of solid fertilizer in the wetland rice field have still not attained on-farm acceptance due to a variety of reasons like, inconsistent performance in terms of high torque requirements, clogging of furrow openers and high power requirement. Some of the machines developed to achieve this objective are prilled urea applicator, Urea Super Granule (USG) applicator, fertilizer injector, etc., which have the operational problems such as the fertilizer discharge is often obstructed because of clogging of furrow openers placed underneath the skids with wet soil in case of prilled urea applicator. The USG applicator [8] required special type of urea formulation in the form of super granules, which have to be prepared through a separate process. This depends upon the availability of the super granules, which has to be done on a large scale. Fertilizer injectors often experienced clogging problems while being applied through contact with the soil.

Precision application of fertilizer is of paramount importance for effective utilization of Nitrogenous fertilizers [9] based on variable-rate application. The problems associated with these fertilizer applicators needed a fresh look to address the problems. A suitable technique needs to be evolved for precise metering and placement of fertilizer with respect to the crop at different stages of growth. Keeping all the above aspects in the view, the present study was undertaken to develop a precision fertilizer applicator to apply precise quantity of nitrogen fertilizer with appropriate depth for wetland rice.

4.2 MATERIALS AND METHODS

Design of the fertilizer applicator needs thorough understanding of various design criteria with due consideration of agronomic requirements of crops like fertilizer application rate and timing of application. In the present study, criteria involve for the design of laboratory simulator and components of the fertilizer applicator were discussed.

4.2.1 ANALYSIS OF FERTILIZER CUP METERING MECHANISM

A fertilizer metering mechanism with a rotor and cup feed type was developed which scoops fertilizer from in primary and secondary compartments and delivers into the delivery spout (Figure 4.1). The metering mechanism in the secondary chamber of fertilizer ensures agitation effect on fertilizer to prevent clogging of fertilizer because of its hygroscopic nature. The

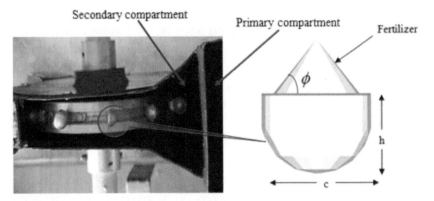

FIGURE 4.1 Fertilizer cup mechanisms inside the fertilizer hopper: c = cup width, h = cup height, φ = angle of repose.

fertilizer from the delivery spout was guided by a V-shaped furrow opener with adjustable depth to place the fertilizer at required depth. Fertilizer metering cup was designed based on the fertilizer requirement of the rice crop. The metering cup was considered as a spherical disk in shape. Dimension of metering cup, cup height (h) and cup width (c) were decided by knowing the application rate of fertilizer (kg/ha).

4.2.2 FERTILIZER HOPPER

Fertilizer hopper design needs prior knowledge of fertilizer properties like bulk density, kinetic angle of repose, particle size, particle strength and hygroscopic behavior. However, some of the researchers have indicated that the depth of fertilizer in the hopper has no effect on flow rate [10]. The bulk density and kinetic angle of repose of urea are the least as compared to other fertilizers like Diammonium Phosphate (DAP) and Muriate of Potash (MOP), which are most widely used in India. The hopper unit was designed to hold 3.0 Kg of granular fertilizer. The granular forms of fertilizers used in this investigation are Urea, DAP and MOP. The material used for the hopper was the mild steel sheet. To overcome the corrosion problem the M.S. sheet was coated with anti-corrosion paint.

4.2.3 FERTILIZER CUP ROTOR

The material used for the design of rotor unit was aluminum to reduce the weight of rotor unit and also to overcome the corrosion problem. The diameter and width of fertilizer metering mechanism rotor were 150 mm and 22 mm, respectively, as shown in Figure 4.2. The number of cups in the circumference of each rotor was 12. The length of the rotor shaft was 120 mm. The inner and outer diameters of rotor shaft were 26 mm and 34 mm, respectively.

4.2.4 DEVELOPMENT OF SIMULATOR FOR FERTILIZER APPLICATOR

The fertilizer flow and distribution characteristics were studied by the design and development of laboratory simulator of fertilizer applicator (Figure 4.3). The developed simulator has provision to vary various design parameters for study under laboratory conditions. The setup comprises a fertilizer hopper, which will be activated by a motor. The fertilizer distributions can be studied at different travel speeds using different metering

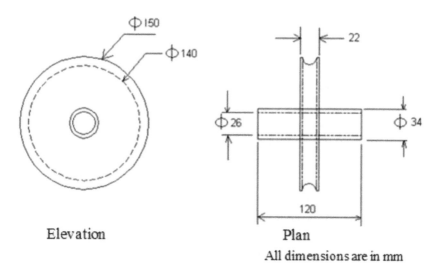

Elevation   Plan

All dimensions are in mm

FIGURE 4.2   Various dimensions of fertilizer rotor unit.

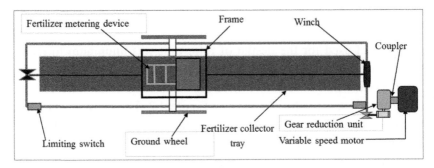

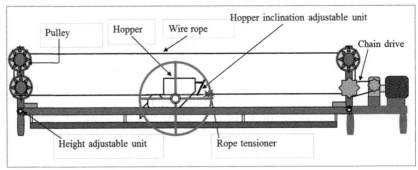

FIGURE 4.3   Laboratory simulator of fertilizer applicator (top and side views).

mechanisms. The mechanism was provided to and fro motion for studying the distribution patterns continuously and with provision to apply fertilizer only during forward motion of the applicator. The data generated is helpful to develop full scale model of the fertilizer applicator.

The total length of the simulator was 6.0 m and the operational length was 4.0 m to take care of inertial effects on the flow of fertilizer during the start of the test run. Two ground wheels supported hopper and metering mechanism unit. The rotor gets the power from ground wheel. Each rotor consists of 12 numbers of cups. The mechanism provided to and fro motion through the variable speed motor. The controller unit controls the speed of operation through variable speed motor. The motor was coupled through a gear reduction unit. Power was transmitted from variable speed motor to gear reduction unit and gear reduction unit to winch mechanism. Winch was connected with three pulley and fertilizer metering mechanism trough wire rope.

FIGURE 4.4   The dog clutch assembly facilitates to rotate ground wheel in one direction.

The experiment was conducted with three different speeds of operation like 1.0, 1.3 and 1.6 km/h, respectively. The limiting switches were used at extreme point of both ends so that when the mechanism comes at the end of the test run its power gets cut off. The tray channel was used to collect the fertilizer. The dog clutch assembly (Figure 4.4) allows the metering mechanism to rotate in one direction only so that the fertilizer drop in to motion and fro motion the ground wheel rotate freely without transmitting power to rotor shaft. The time taken to cover operational distance (4.0 m) was recorded.

4.3   RESULTS AND DISCUSSION

The simulator was evaluated on some of the commercially available fertilizers though the basic objective of this is to use the implement for applying Urea in different stages of crop growth. However, the machine can be effectively used for applying other fertilizers like Phosphatic and Potassic fertilizers, which are applied usually in single application at basal stage.

4.3.1   PROPERTIES OF FERTILIZERS

The properties of fertilizers were analyzed (Table 4.1). Sieve analysis was done to characterize their size distribution. The DAP is more granular in form with size in the range of 2.28 to 4.26 mm followed by Urea (2.02 to 2.49 mm) and MOP which is anhydrous in nature. Urea has very low

TABLE 4.1   Properties of Urea, DAP and MOP Fertilizers Used for Present Study

Fertilizer	Bulk Density (kg-m⁻³)	Angle of Repose, degrees	Size range, mm	Moisture content, % db
DAP	931.1	31.26	2.28 to 4.26	1.01
MOP	1086.5	33.69	<1.0	17.64
Urea	764.4	29.98	2.02 to 2.49	1.52

density of 764 kg/m³ as compared to MOP (1086 kg/m³). The moisture content at which the studies were carried out was also mentioned.

4.3.2 EFFECT OF TRAVEL SPEED ON APPLICATION RATE

There is a chance of backflow or the falling of fertilizer sideways while operating at different speeds. The three fertilizers were applied at travel speeds of 1, 1.3, and 1.6 km/h. The nature of fertilizers had a fair effect on the application rate (Figure 4.5). From the statistical analysis (Table 4.2)

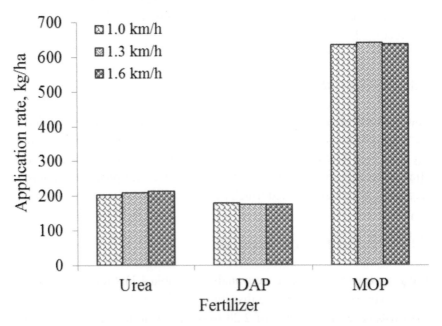

FIGURE 4.5   Effect of travel speed on application rate.

TABLE 4.2   One-Way Analysis of Variance Test on Uniformity in Application Rate

Source of Variation	SS	df	MS	F	Fcrit	Grand mean
Urea						
Treatments	7.7	2	3.9	0.09	10.9	210.0
Error	264.4	6	44.1			
Total	272.1	8				
CV[a]	3.2					
DAP						
Treatments	2.74	2	1.37	0.09	10.9	177.2
Error	94.5	6	15.75			
Total	97.2	8				
CV[a]	2.2					
MOP						
Treatments	6.16	2	3.08	0.07	10.92	637.83
Error	247.40	6	41.23			
Total	253.56	8				
CV[a]	1.0					

[a] CV = Coefficient of Variation.

on the variation in application rates at different travel speeds (which are considered as treatments). It is very clear that the effect of travel speed is highly insignificant (at 1% level of confidence with CV being very low in the range of 1–3.2%). It confirms the fact that the metering mechanism is working perfectly without any backflow or sidewise falling of fertilizer.

The cup with dimensions of 16×6 mm resulted in Urea application in the range of 203 to 216 kg/ha, DAP in the range of 175 to 179 kg/ha and MOP in the range of 636 to 647 kg/ha at different travel speeds. Since the design is aimed at urea, the cup design obviously is giving higher application rates with other fertilizers. The cup dimensions need to be redesigned for these.

4.3.3   UNIFORMITY OF FERTILIZER DISTRIBUTION

The samples were collected (Table 4.3) in a forward travel stretch of 20 cm randomly and analyzed for their distribution. The CV was higher in

TABLE 4.3   Uniformity of Fertilizer Distribution

Param-eters	Urea			DAP			MOP		
Travel speed	1.0	1.3	1.6	1.0	1.3	1.6	1.0	1.3	1.6
Mean	197.5	196.3	201.3	222.2	245	258.8	591.3	591.3	606.3
SD	21.1	18.7	19.0	49.1	23.0	37.8	38.2	89.4	49.7
Range	6.5	15.1	8.2	62.5	50.0	50.0	150.0	37.5	112.5
CV	10.7	9.4	9.5	22.1	9.4	14.6	6.5	15.1	8.2

the lowest travel speed tested, which may be due to possible distortion in distribution on ground at lower speeds. Travel speed of 1.3 km/h was found to be optimal.

The studies reveal that the fertilizer distribution is varying within the range of 9.4 to 22.4%. The higher CV at slower speed may be attributed to the higher time for the fertilizer to reach ground surface thereby affecting its uniformity of distribution.

4.3.4   MODELING OF FERTILIZER APPLICATOR

A computer program was developed to evaluate cup height for a rotor of given width to select suitable size for different fertilizers at different application rates. The program based on the flow chart represented by Figure 4.6 (where, c = cup width, h = cup height, ϕ = angle of repose, D = diameter of ground wheel, ρ = density of fertilizer, n = number of splits fertilizer requirement, Q = nutrient requirement, b = width of coverage, etc.) has provision for split applications of fertilizer also. The cup height evaluated on iteration process till the values converge in the range of $\leq 1 \times 10^{-8}$. The cup height varied from 4 to 7 mm for applying MOP, 3.5 to 6 mm for DAP and 3.2 to 6.7 mm for Urea application in rice. Suitable cup height for a given application can be chosen from the graph for different fertilizers (Figure 4.7). The cup height refers to a base cup width of 16 mm, which is decided in relation to the rotor width.

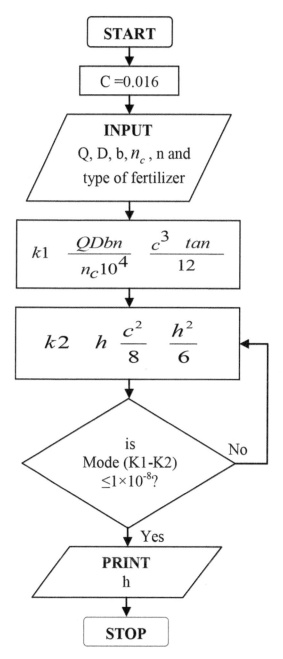

FIGURE 4.6 Flow chart for evaluating different application rates though changing cup dimensions.

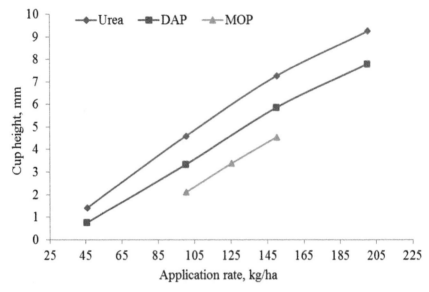

FIGURE 4.7   Variation of cup height with different application rates of fertilizers.

4.4   SUMMARY

Most of the available precision fertilizer applicators are expensive and difficult to purchase by small and marginal farmers. The fertilizer applicator of cup feed mechanism described in paper has a simple structure, easy to use, and low price, suitable for the vast rural areas. The fertilizer applicator has provided with dog clutch mechanism that facilitates discharging of fertilizer during forward travel only for minimizing the loss of fertilizer. The simulator was used for testing fertilizer applicator for commercially available fertilizers like Urea (Nitrogenous), Diammonium Phosphate (DAP) (Phosphatic) and Muriate of Potash (MOP) (Potassic). The performance was evaluated considering the uniformity in distribution and application rate. The cup feed mechanism was effective in giving uniform distribution and also suitable for different fertilizers as the cups are replaceable. The program developed was useful in selection of cup size depending on nutritional requirement of the crop. Laboratory trials have shown that the applicator can do this successfully with urea, DAP and MOP. The effect of travel speed had insignificant influence on application rate of fertilizers. The fertilizer was uniformly distributed along the row within the range of 9.4 to 22.4%.

KEYWORDS

- **application rate**
- **DAP**
- **fertilizer applicator**
- **fertilizer nutritional requirement**
- **granular fertilizer**
- **MOP**
- **precision fertilizer applicator**
- **simulator**
- **Uniformity**
- **Urea**

REFERENCES

1. Fertilizer Association of India (2010). Fertilizer Statistics 2009–2010 and earlier issues. The Fertilizer Association of India, New Delhi.
2. Youngdahl, L. J., Lupin, M. S., & Craswell, E. T. (1986). New developments in nitrogen fertilizers for rice. Fertilizer Research, 9(1–2), 149–160.
3. Malhi, S. S., & Nyborg, M. (1992). Recovery of nitrogen by spring barley from ammonium nitrate, urea and sulfur-coated urea as affected by time and method of application. Fertilizer Research, 32(1), 19–25.
4. Stevens, R. J., & Laughlin, R. J. (1989). A microplot study of the fate of 15N-labeled ammonium nitrate and urea applied at two rates to ryegrass in spring. Fertilizer Research, 20(1), 33–39.
5. Strong, W. M., Saffigna, P. G., Cooper, J. E., & Cogle, A. L. (1992). Application of anhydrous ammonia or urea during the fallow period for winter cereals on the Darling Downs, Queensland. II. The recovery of 15N by wheat and sorghum in soil and plant at harvest. Soil Research, 30(5), 711–721.
6. Diekmann, K. H., De Datta, S. K., & Ottow, J. C. G. (1993). Nitrogen uptake and recovery from urea and green manure in lowland rice measured by 15N and non-isotope techniques. Plant and Soil, 148(1), 91–99.
7. Bautista, E. U., Koike, M., & Suministrado, D. C. (2001). Mechanical Deep Placement of Nitrogen in Wetland Rice. Journal of Agricultural Engineering Research, 78(4), 333–346.
8. Khan, A. U. (1984). Deep placement fertilizer applicator for improvement fertilizer use efficiency. Agricultural Mechanization in Asia, Africa and Latin America, 15(3), 25–32.

9. Umeda, M., Kaho, T., Lida, M., & Lee, C. K. (2001). Effect of variable rate fertilizer for paddy field. ASAE Paper No. 011111. ASABE, St Joseph – MI.
10. Oni, K. C. (1982). Performance evaluation of a modified fertilizer applicator. Agricultural Mechanization in Asia, Africa and Latin America, pp. 67–70.

APPENDIX I
AVAILABLE FERTILIZER APPLICATORS

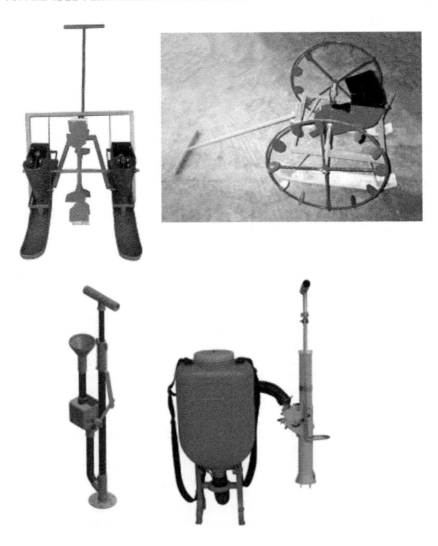

PART 2

AGRICULTURAL PROCESS
ENGINEERING

CHAPTER 5

HYPOBARIC STORAGE OF HORTICULTURAL PRODUCTS: A REVIEW

P. VITHU and J. A. MOSES

CONTENTS

5.1 INTRODUCTION

Fruits and vegetables have high nutraceutical value and are important constituents of our daily diet. In general, horticultural products are highly perishable and require appropriate preservation techniques to maintain their quality and remain fresh. As far as these perishables are concerned, minimizing post-harvest losses is a major challenge. FAO estimates that post-harvest losses in fruits, vegetables and root crops are almost 40–50%

[2]. Losses occur during harvesting, handling, transporting and storage. Adopting proper storage techniques and focusing on food processing and preservation are proven remedies to minimize post-harvest losses.

Safe storage refers to the holding of fresh fruits and vegetables, protected from deterioration, for extended periods. Storage helps to meet continuous market demands, contributes in price stabilization, and provides year-round availability of various off-season and exotic fruits and vegetables. However, 'safe' storage of perishables requires cautious control over several factors that affect storage quality and shelf life. These include crop, harvest, handling, transportation and storage conditions. In the context of storage conditions, important considerations include: storage temperature, relative humidity, gas compositions and interactions with the environment.

Heavy losses have been reported in traditional storage structures. Several improved and novel packaging and storage technologies have been experimented/commercialized for fruits and vegetables. These include refrigerated storage, modified atmosphere packaging, intelligent packaging, nano-packaging, edible packaging, controlled atmospheric storage, silicone membrane technology, hypobaric storage and vacuum packaging. Important considerations for commercially adopting these technologies for bulk-level include: cost effectiveness, eco-friendliness, product quality and applicability range, storage area– to-product volume ratio, skill requirement, process control, energy requirement and safety aspects.

Hypobaric pressure or sub-atmospheric pressure refers to pressure below 101 kPa. Hypobaric storage is a relatively less understood type of controlled atmospheric storage in which low-oxygen environment is created at sub-atmospheric pressure ranges; which in turn reduces the respiration rates and metabolism kinetics of commodities, thereby increasing storage life [39]. Reduction in total pressure is proportional to reduction in the partial pressure of oxygen [24]. The produce is stored under partial vacuum in a chamber which is vented continuously with saturated air, to maintain such low-oxygen partial pressure ranges. Generally, a reduction in air pressure of 10 kPa (equivalent to an oxygen partial pressure of 2.1 kPa) is permits 2% reduction in oxygen concentration at normal atmospheric pressure.

The concept, that storage atmosphere pressure affects produce metabolism, was first observed by Workman et al. [40] during studies conducted on tomato fruits. They concluded that storage under increased pressure increased the respiration rate of tomatoes. However, in-depth understanding of this pressure-metabolism relationship had its origin in a study conducted for determining the equation governing gas exchange in fruits [15]. This equation explains that the rate of gas passage in the outward direction equals the rate at which gas is produced within the commodity [11]. Further, the partial pressure of individual gas component in the air has a direct relationship with the total pressure of the system. Thus, storage under low pressure can reduce oxygen availability to the produce. Almost half-century research has been conducted on hypobaric storage for fruits and vegetables. In spite of several positive aspects, the technology remains non-commercialized.

This chapter includes the underlying principle of hypobaric storage, various designs of hypobaric systems and the effect of low-pressure storage on the physiological and biochemical quality of food products. Recent trends in this concept and directions for future work are also presented. A review is presented on hypobaric storage technology of horticultural products.

5.2 WORKING PRINCIPLE OF A HYPOBARIC STORAGE SYSTEM

A typical hypobaric system consists of a produce storage chamber, a vacuum pump, a refrigeration unit and a humidifier (Figure 5.1). Air enters the system through the airflow chamber (1). A needle valve (2) is used to regulate the incoming low pressure air [whose downstream pressure is measured using a vacuum gauge (3)]. Air is allowed to enter a humidifier (4) to increase the relative humidity of the air to about 80–100% (as in most cases). For this purpose, distilled water (5) from a water reservoir (6) is added periodically to the humidifier, with the help of a valve (7). The humidification system removes respiratory heat and any other additional heat received from the environment; thereby preventing water loss. Saturated air is supplied to the vacuum storage chamber (8) through a conduit (9). Food commodities are stored in this chamber under vacuum

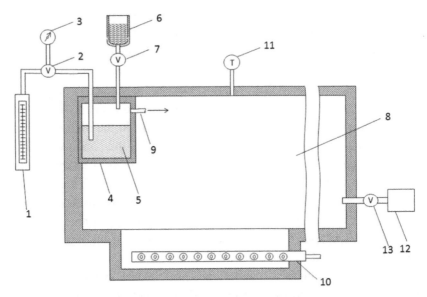

FIGURE 5.1   Schematic diagram of a hypobaric storage unit [12]: (1) Airflow chamber, (2) needle valve, (3) vacuum gauge, (4) humidifier, (5) distilled water, (6) water reservoir, (7) valve, (8) storage chamber, (9) conduit, (10) refrigeration unit, (11) temperature gauge, (12) vacuum pump, (13) throttle valve.

(usually 4–400 mm Hg absolute). Storage temperature is an important parameter and is kept under control (–2 to 15°C) using a refrigeration unit (10) designed in the form of coolant tubes. A temperature gauge (11) shows the variation in temperature during storage. To ensure maintaining low pressure in the storage chamber, a vacuum pump (12) provided. A pressure regulator/throttle valve (13) controls the entry/exit of air from/ to the system.

Burg and Burg [14] found that for all temperature ranges used for hypobaric storage, temperature correction was insignificant and therefore, gas diffusivity of the commodity is inversely related to absolute pressure. Reduced pressure can promote proliferation of gaseous components from within the plant tissue and accelerate diffusion kinetics of ethylene. When the commodity is transferred from hypobaric to atmospheric conditions, it resumes emanation of volatiles at a normal rate [18]. According to Burg [10], if a commodity in equilibrium with its environment is transferred from atmospheric pressure to rarefied air having an absolute pressure of 0.1 atm, when a new equilibrium is established, volatiles will be produced and escape at the same (initial) rate. However, the concentration gradient driving out the gas adjusts to a lower value (1/10th of the initial), because of a ten-fold increase in diffusivity.

Hypobaric storage rapidly removes respiratory heat, and any other heat received from the environment. Under reduced pressure, heat produced by respiration is transferred by evaporative cooling and water vapor can be evacuated without refrigeration [16]. Products can be densely packed in a hypobaric storage chamber and thus be able to maintain a more uniform temperature, humidity and gas composition, favorable for storage. Products can therefore be stored for longer duration in fresh condition.

5.3   DESIGN OF A HYPOBARIC STORAGE SYSTEM

Broadly, hypobaric system can be classified as 'wet' or 'dry' types. Both these methods use the same equipment, except a device which controls the pumping speed in the case of 'dry' hypobaric method, in place of the humidifier in the 'wet' hypobaric method. Another classification of hypobaric system is based on the pressure treatment (regular suction or static type and continuous suction or dynamic type). Hypobaric systems can be built as warehouses or intermodal containers. A hypobaric warehouse is a sealed commercial building with an inbuilt hypobaric system to store fresh produce. On the other hand, low pressure intermodal containers are used for transporting fresh fruits and vegetables. The unit consists of a pressure controlled module and a refrigeration unit. Design variants of hypobaric warehouses and low pressure intermodal containers are given in Tables 5.1 and 5.2, respectively. Generally, in both warehouses and intermodal containers, storage boxes are used to store fruits and vegetables. Theses boxes are designed sufficiently to maintain box stacking strength and crush resistance, even under high humidity storage conditions. Therefore, storage boxes are laminated with water- resistant materials. Materials such as paperboard boxes, fiber box approved cardboard boxes (like those impregnated with wax coating) or vented open-topped water-proof plastic bins have been used for storage purposes [9].

Hypobaric storage is closely associated with commercially adopted controlled atmosphere storage techniques. With growing interests on controlled atmosphere storage, the feasibility of adopting large-scale, commercial hypobaric storage systems are now explored. In spite of numerous benefits and advantages of hypobaric storage compared with the controlled atmospheric storage [10], it has not gained much acceptance in commercial storage of fruits and vegetables. Research is needed to develop portable or palette-sized systems without generating the prohibitive cost factor.

5.4   EFFECT OF HYPOBARIC STORAGE ON FOOD QUALITY

'Quality' of a food product is the 'fitness to use'; meeting the expectations of the consumer. During storage, foods may be subjected to several complex physiological and biochemical changes which affect its quality. Fruit metabolism, decay rate, respiratory parameters, ethylene production, enzyme activity and several other bio-chemical pathways are important parameters deciding the quality of stored products.

Under atmosphere pressure, fermentation and associated damages are important issues. But under hypobaric storage, with (up to 100-fold) lower oxygen levels, better quality produce can be obtained. It also preserves the tissue better than most other methods of storing horticultural products. Hypobaric storage prevents ethylene-induced responses, low oxygen and high carbon dioxide injuries, inhibits bacterial and fungal growth, can vacuum-fumigate the commodity with HOCl vapor, and has the potential to kill surface insects [9]. The technique is known to extend the shelf-life of fresh fruits, vegetables, cut-flowers, potted plants, meat, poultry, fish, shrimp, and other metabolically active matter [12]. Also, low pressure can kill all life stages of several insects of agricultural products [13], mainly because of low oxygen concentrations. Therefore, it could permit its usage as an alternative to several chemical based disinfestations techniques; which are a matter of global concern.

Hypobaric storage is recognized as a safe approach that preserves the freshness of food [27]. The effects of hypobaric storage on physiology and quality of fresh produce have been studied extensively [19, 28, 41]. Many studies have been reported so far, on the potential of hypobaric storage for

TABLE 5.1   Design Variants of Hypobaric Warehouses [13]

Name	Dimensions	Construction material	Specifications	Capacity
Alternate steel cylinder	3.35 m diameter, 19.8 m high cylinder	Steel	Saddle mounted carbon steel or stainless steel vacuum tank; jacketed refrigeration system	676.58 m³
Boeing or Preload warehouse	Inside cylinder: 22.9 m diameter and 6.7 m high; outside cylinder: 14 m diameter and 7.6 m high	Concrete	Two cylindrical tanks, vacuum: 10.1 kPa; temperature range: 0–26°C; forced-air refrigeration system	1019 m³ (inside)
Concrete masonry corporation	18.3 m diameter, 7.6 m high cylinder	Concrete; metal frame for pressure vessel door	Roof pieces rest on reinforced ring crown	1982 m³
Crop corporation warehouse	10.7 m diameter, 7.6 m high cylinder	Concrete; polyurethane coating at roof dome and asphalt neoprene coating at wall exterior	Circular wire prestressed composite structure; dome shaped structure with thin concrete shell	706 m³
Grummans's hypobaric warehouse	3.35 m diameter, 19.8 m high cylinder	Polyurethane insulated steel	Temperature range: 0–18.3°C; relative humidity range: 90–95%; absolute pressure range: 1.33–13.3 kPa	676.58 m³
University of Guelph design	8.5 m diameter, 4.9 m high cylinder	Reinforced cement concrete	4 compartments of equal capacity; internal insulation and external coating to prevent air leakage	Total capacity: 343 m³; compartment capacity: 86 m³

delaying ripening and increasing shelf life of fresh fruits. Transformation of 1-aminocyclopropane carboxylic acid to ethylene involves an oxidation reaction. Therefore, low oxygen partial pressure under hypobaric conditions reduces the rate of ethylene production and therefore delays fruit

TABLE 5.2   Designs Developed for Hypobaric Intermodal Containers [13]

Name	Dimensions	Features	Capacity
Fruehauf Corporation Prototype No. 1	1.8 m cylindrical stainless steel tank (inside); insulated container 6.1 x 2.6 x 2.4 m (outside)	Standard forced air refrigeration unit; cold jacket prevent heat infiltration; polyurethane insulation	38.064 m³
Fruehauf Corporation Prototype No. 2	2.44 m x 2.44 m x 12.2 m vacuum tank	Aluminum vacuum tank inside non-insulated tank; tank's reinforcement alternate with cylinder's strut ribs	73 m³
Grumman Domavac	12 m x 12 m x 0.203 m size double walled panel	Inner walls with polyurethane insulation; no metal to metal contacts between inner and outer walls	42.45 m³
VacuFresh Tank	6.1 m x 2.6 m x 2.4 m	Relative humidity range: 0–100%; temperature range: -17.8 to 49.2°C; self-supporting cylindrical container; designed to withstand full internal vacuum	21.2 m³
VivaFresh	2.5 m x 3 m x 12.2 m	Pressure range: 10–20 kPa; temperature range: 0–10°C; relative humidity range: 95–100%	94 m³

ripening [6]. Further, oxygen levels of 0.06 to 0.15% restrict the onset of fermentation [42]. Apelbaum et al. [3] studied the effect of low pressure (13.33 kPa and 6.66 kPa at 13°C) storage on the quality of mango fruit and observed 9 and 19 days delay in ripening, respectively. 'Lula' avocados remained firm for over 3 months under hypobaric conditions and then ripened normally after the storage period. This extended shelf life under low pressure is almost twice of that reported when avocados were stored under conventional controlled atmosphere [33].

Under hypobaric storage, the produce undergoes a series of changes, both favorable and undesirable. These depend on the type of food product and storage conditions, as physiological and biochemical factors depend on storage pressure, temperature, relative humidity and storage time. Studies on different varieties of fruits and vegetables showed that under hypobaric conditions, reduction in firmness and chlorophyll content was lower, abscissic acid synthesis was prevented, ascorbic acid content was increased, acid catabolism was reduced, sugar was maintained at a higher value, electrolyte leakage and physiological disorders such as internal breakdown were reduced or eliminated, shelf-life after transfer to ambient conditions was greatly improved, and respiration, ethylene synthesis and ethylene content were remarkably lowered [4, 8, 19]. Table 5.3 summarizes the effect of hypobaric storage on selected fruits and vegetables. There is a need to further investigate the effect of short-term and long-term hypobaric treatment on fruit physiology.

Products stored under hypobaric conditions may suffer weight loss, and therefore controlling relative humidity is critical [12]. Laurin et al. [26] from their study on cucumber found that hypobaric storage (at 70.9 kPa for 6 hours at 20°C) causes weight loss in fresh produce (significantly more than that under atmosphere storage). Further, it was observed that stomata of low-pressure treated cucumbers remained open even after 96 h after returning to ambient conditions. Delay in stomata closure due to stress resulted in increased weight loss. Use of humid air can reduce this effect [43]. One of the other benefits of hypobaric storage of fresh produce is the reduction in fruit browning and flesh leatheriness as observed in studies conducted on loquat [20].

Further, short-term low-pressure treatments show positive results for the control of different fungal rots and decay. Romamazzi et al. [31] conducted studies on the effect of short-term hypobaric storage on postharvest fungal rot of strawberries, sweet cherries and table grapes. Best results for reduction in fungal decay (due to *R. stolonifer* and *B. cinerea*) were obtained under 25 and 50 kPa treatments in sweet cherries. Hashmi et al. [23] demonstrated that hypobaric treatment (50 kPa, 4 h) reduced fungal rots in blueberries stored at 20°C for 7 days, with no effect on physiological weight loss, firmness and respiration rate.

Generally hypobaric storage decreases respiration rate and ethylene production. However, Plaxton and Podesta [30] reported that hypobaric storage increases respiration rate. This is attributed to the stress responses of the fruit. Similar results were observed by Tovar et al. [35] and hence, suggest the need for in-depth research on the effect of hypobaric storage on various bio-chemical changes in the fruit.

TABLE 5.3   Effect of Hypobaric Storage on the Quality of Selected Fruits and Vegetables

Commodity	Variety	Storage conditions	Physiological/biochemical changes during storage	Ref.
Avocado	Choquette	5.3–13.3 kPa at 14.4°C for 17 days	Firm with no external blemishes; normal taste when transferred to ambient conditions, even after 3 days of storage	[11]
Cabbage	Summer	7.33–8.0 kPa at 1°C and 1.4–1.5% oxygen for 42 days of storage	Greener and better appearance; 2.4% weight loss; no off-odors	[29]
Cress	Garden	10 kPa at 3°C for 12 days	Retained ascorbic acid, protein and chlorophyll levels	[4, 5]
Cucumber	Golden boy	10 kPa and 100 kPa at 10°C for 30 days	Reduced dietary fiber drop; minimal increase in pectin content; decrease in total sugar content (1.11 g/100 g to 1.26 g/100 g and 1.95 g/100 g samples, at 10 kPa and 100 kPa, respectively)	[43]
Dry onion	Autumn spice; rocket; trapps	8 kPa at 26°C and 56% relative humidity	12.2% weight loss; over-cured onions	[29]
Jujube	Lizao, Dongzao	20 Pa-50 kPa	Decrease in firmness and titrable acidity; 2.5 % weight loss; increase in malondialdehyde content, polyphenol oxidase activity and browning rate	[17]
Radish	Topped spring	10 kPa at 2.2–2.8°C and 95% relative humidity	Insignificant loss in chlorophyll; reduction in protein and ascorbic acid contents	[4]
Spinach	Green	10 kPa at 95 % relative humidity for 7 weeks	Retention of ascorbic acid, green color and total protein	[4]
Squash	Acorn	7.33–8.0 kPa at 2% oxygen and 90–95% relative humidity	4.2% weight loss	[29]

5.5 RECENT TRENDS IN HYPOBARIC STORAGE

Wen-Xiang and Min [37] conducted a study on hypobaric storage in combination with atmospheric cold storage. It was found that the storage life of honey peach was better in terms of freshness, taste and flavor, color, and texture. These peaches could be stored for lesser days when kept in cold stores (only). The theoretical basis for this combined effect must be further explored. Similarly, Zhang et al. [44] conducted studies on the effects of low temperature, hypobaric and ozone treatments on the physiological and biochemical attributes of Dong Jujube fruit. Results showed that hypobaric and ozone treatments reduced the rate of respiration, inhibited amylase and ascorbic acid oxidase activity, retarded starch and ascorbic acid degradation rates, inhibited mold reproduction, reduced fruit rot and maintained fruit firmness. Application of hypobaric storage reduced chilling injury of Dong Jujube fruit and, thus lower storage temperatures could be used.

Hypobaric storage in combination with low temperature is a feasible method for extending the shelf-life of fruits and vegetables. Wang and Zhang [36] studied the effects of hypobaric treatment on physiological and biochemical attributes of apricots stored under low pressure. Results showed that fruit firmness under low-pressure storage was always higher than when stored under atmospheric pressure. Hypobaric treatment could delay softening and senescence of apricots by reducing respiration rate, polygalacturonase and lipoxygenase activities and maintaining fruit structure. During respiration, fresh produce give off carbon dioxide, water and heat. According to Van't Hoff's rule, respiration rate doubles for every $100°C$ increase in temperature.

A study was conducted by Wen-Xiang and Min [37] to understand the effect of hypobaric conditions on the quality changes during storage. They studied the effect of a three-stage hypobaric storage condition and atmosphere cold storage on the quality and senescence of asparagus. In the three – stage hypobaric storage method, pressure is increased periodically, as three different phases during the storage. The initial phase uses low pressure (15–20 kPa for 3 days), followed by second (25–30 kPa for 1 week) and third (45–50 kPa till the end of storage) phases. The three-stage hypobaric storage method significantly inhibited increase in cellulose, lignin, water soluble pectin contents, shear and puncture intensity, and the degradation rate of protopectin. This study suggested the scope of development of three-stage hypobaric storage systems for other food storage applications.

The food logistics industry seeks for innovative technologies for safe and efficient transport of products. Hypobaric intermodal containers have been proved to be effective even over long distances. A hypobaric storage intermodal container is a complex structure which can attain the low pressure condition in few hours. Operating the unit is simple; venting takes few minutes and working personnel can enter the unit immediately thereafter. Mechanical pressure regulators and flow controllers facilitate pressure and relative humidity control.

5.6 CONCLUSIONS

Though hypobaric storage systems are expensive, recent studies suggest that the technique can be more effective when used in combination with other technologies. Different modes of application of low pressure are also under study. However, further research is required for defining management protocols for safe and extended storage of perishables under low-pressure conditions.

5.7 SUMMARY

Hypobaric or low-pressure storage is an advanced method for safe storage of perishables. A controlled atmosphere is created by reduced pressure which in turn lowers partial pressure of oxygen in the storage chamber. The technique is known to extend the shelf-life of fresh fruits, vegetables, cut-flowers, potted plants, meat, poultry, fish, shrimp, and other metabolically active matter. This paper covers the underlying principle of hypobaric storage, various designs of hypobaric systems and the effect of low-pressure storage on the physiological and biochemical quality of food products. Recent trends in this concept and directions for future work are also presented.

KEYWORDS

- food quality
- hypobaric storage
- low pressure storage
- perishables
- shelf life

REFERENCES

1. Acedo, A. L. Jr., & Weinberger K. (eds.) (2009). Proceedings of the GMS workshop on economic analysis of postharvest technologies for vegetables. 19–21 August 2008, Siem Reap, Cambodia. AVRDC Publication No. 09–730. AVRDC – The World Vegetable Center, Taiwan. 117 p.
2. Anon. (2013). Accessed on 19.08.2015 from: http://www.fao.org/in-action/seeking-end-to-loss-and-waste-of-food-along-production-chain/en/
3. Apelbaum, A., Zauberman, G. and Fuchs, Y. (1977). Prolonging the storage life of avocado fruits by sub-atmospheric pressure. Horticultural Science, 12(2), 113–117.
4. Bangerth, F. (1973). The effect of hypobaric storage on quality, physiology, and storage life of fruits, vegetables and cut flower. Gartenbauwissenschaft, 38, 479–508.
5. Bangerth, F. (1974). Hypobaric storage of vegetables. Acta Horticultarae, 1(38), 2332.
6. Beaudry, R. M. (1999). Effect of O2 and CO2 partial pressure on selected phenomena affecting fruit and vegetable quality. Postharvest Biology and Technology, 15(3), 293–303.
7. Beaudry, R. M. (2009). Future trends and innovations in controlled atmosphere storage and modified atmosphere packaging technologies. In: X International Controlled and Modified Atmosphere Research Conference, 876, 21–28.
8. Bérard, L. S., & Lougheed, E. C. (1982). Electrolyte leakage from daminozide-treated apples Malus-Domestica cultivar McIntosh held in air, low pressure, and controlled atmosphere storage. Journal of the American Society of Horticultural Science, 107, 421–425.
9. Burg, S. P. (2014). U.S. Patent No. 8,763,412. Washington, DC: U.S. Patent and Trademark Office.
10. Burg, S. P. (1967). Method for storing fruit. US Patent 3,333,967 and US Patent Reissue Re., 28, 995.
11. Burg, S.P. (1968). Ethylene, plant senescence and abscission. Plant Physiology, 43, 1503–1511.

12. Burg, S. P. (1976). Low temperature hypobaric storage of metabolically active matter. US patents 3,958,028 and 4,061,483.
13. Burg, S. P. (2004). Post-harvest physiology and hypobaric storage of fresh produce. CAB International, Wallingford, Oxfordshire, UK, p654.
14. Burg, S. P., & Burg, E. A. (1967). Molecular requirements for the biological activity of ethylene. Plant Physiology, 42, 144–152.
15. Burg, S. P., Burg, E. A. (1966). Fruit storage at sub-atmospheric pressures. Science, 152, 314–3.
16. Burg, S. P., & Kosson, R. L. (1983). Metabolism, heat transfer and water loss under hypobaric conditions. In: Post-Harvest Physiology and Crop Preservation. Lieberman, M. (Ed.), Plenum Press, New York, NY, 399–424.
17. Chang, Y. (2002). The developing and research foreground on new technology of hypobaric storage. Machinery for Cereals Oil Food Process, 2, 8–9.
18. Dilley, D. R. (1977). The hypobaric concept for controlled atmosphere storage. In: Controlled Atmospheres for the Storage and Transport of Perishable Agricultural Commodities. Dewey, D. H. (Ed.), Proceedings of the Second National CA Research Conference, April 57, 1977. Michigan State University, 29–37.
19. Dilley, D. R. (1978). Approaches to maintenance of postharvest integrity. Journal of Food Biochemistry, 2, 235–242.
20. Gao, H. Y., Chen, H. J., Chen, W. X., Yang, J. T., Song, L. L., & Jiang, Y. M. (2006). Effect of hypobaric storage on physiological and quality attributes of loquat fruit at low temperature. Acta Horticulturae, 712, 269–274.
21. Haard, N. F., & Lee, Y. Z. (1982). Hypobaric storage of Atlantic salmon in a carbon dioxide atmosphere. Canadian Institute for Food Science and Technology Journal, 15, 6871.
22. Haiyan, G., Lili, S., Yongjun, Z., Ying, Y., Wenxuan, C., Hangjun, C., and Yonghua, Z. (2008). Effects of hypobaric storage on quality and flesh leatheriness of cold-stored loquat fruit. Chinese Society of Agricultural Engineering, 24(6), 245–249.
23. Hashmi, M. S., East, A. R., Palmer, J. S., & Heyes, J. A. (2013). Hypobaric treatment stimulates defense-related enzymes in strawberry. Postharvest Biology and Technology, 85, 7782.
24. John, J. (2008). A handbook on post-harvest management of fruits and vegetables. Daya publishing house, Delhi. p.119.
25. Kumar, P. S., & Meena, U. S. (2005). Glimpses Of Post-Harvest Technology. New Vishal Publications, 47–59.
26. Laurin, É., Nunes, M. C. N., Émond, J.P and Brecht, J. K. (2006). Residual effect of low pressure stress during simulated air transport on Beit Alpha-type cucumbers: stomata behavior. Post-harvest Biology and Technology, 41(2), 121–127.
27. Liu, Y.-B. (2003). Effects of controlled atmosphere treatments on insect mortality and lettuce quality. Journal of Economic Entomology, 96(4), 1100–1107.
28. Lougheed, E. C., Murr, D. P., & Bérard, L. (1978). Low-pressure storage of horticultural crops. Horticultural Science, 13(1), 2127.
29. McKeown, A. W., & Lougheed, E. C. (1980). Low-pressure storage of some vegetables. Acta Horticulturae, 116, 83–100.
30. Plaxton, W. C., & Podesta, F. E. (2006). The functional organization and control of plant respiration. Critical Reviews in Plant Sciences, 25(2), 159–198.

31. Romanazzi, G., Nigro, F., Ippolito, A., Salerno, M. (2001). Effect of short hypobaric treatments on postharvest rots of sweet cherries, strawberries and table grapes. Postharvest Biology and Technology, 22, 16.

32. Shirke, P. S. (2007). Post-Harvest Engineering of Fruits and Vegetables. CBS Publishers and Distributors, New Delhi, 75–91.

33. Spalding, D. H., & Reeder, W. F. (1976). Low pressure (hypobaric) storage of avocados. Horticultural Science, 11, 491–492.

34. Stenvers, N., & Bruinsma, J. (1975). Ripening of tomato fruits at reduced atmospheric and partial oxygen pressures. Nature, 253, 532–533.

35. Tovar, B., Montalvo, E., Damian, B. M., Garcia, H. S., & Mata, M. (2011). Application of vacuum and exogenous ethylene on ataulfo mango ripening. LWF-Food Science and Technology, 44(10), 2040–2046.

36. Wang, W., & Zhang, Y. (2008). Control of hypobaric treatment on softening of postharvest apricot fruits. Acta Botanica Boreali Occidentalia Sinica.

37. Wen-xiang, L., & Min, Z. (2005). Effects of combined hypobaric and atmosphere cold storage on the preservation of honey peach. International Agrophysics, 19(3), 231.

38. Wen-Xiang, L., Zhang, N., & Hab-qing, Y. (2006). Study on hypobaric storage of green asparagus. Journal of Food Engineering. 73, 225–230.

39. Wills, R. B. H., McGlasson, W. B., Graham, D., & Joyce, D. C. (2007). Post-harvest: An introduction to physiology and handling of fruits, vegetables and ornamentals, 5th edition, 91.

40. Workman, M., Pratt, H. K., & Morris, L. L. (1957. Studies on the physiology of tomato fruits. I: Respiration and ripening behavior at 20°C as related to date of harvest. Journal of the American Society of Horticultural Science. 69, 352–365.

41. Wu, M. T., Jadhav, S. J., & Salunkhe, D. K. (1972). Effects of subatmospheric pressure storage on ripening of tomato fruits. Journal of Food Science. 37, 952–956.

42. Yeoshua, S. B. (2005). Environmentally friendly technologies for agricultural produce quality. CRC Press. pp. 72.

43. Zapotoczny, P., & Markowski, M. (2014). Influence of hypobaric storage on the quality of greenhouse cucumbers. Bulgarian Journal of Agricultural Science, 20(6), 1406–1412.

44. Zhang, Y. L., Han, J. Q., & Zhang, R. G. (2005). Study on fresh-keeping physiological activity of dong jujube using low temperature combined with hypobaric and ozone treatments. Scientia Agricultura Sinica, 38(10), 2102–2110.

CHAPTER 6

CONVECTIVE DRYING OF POTATO SLICES AT LOW-TEMPERATURES IN FULL AIR RECIRCULATION DRYER

IVAN ZLATANOVIĆ, DUŠAN RADOJIČIĆ, and GORAN TOPISIROVIĆ

CONTENTS

6.1   INTRODUCTION

In modern industrial production, it can be said with certainty that there is no such a product whose main raw material in one of its stages did not go through a drying process [12, 13]. Research and analysis of the drying process is essential for its further optimization and improvement. A clear physical and mathematical description of the process of drying biological material is still pretty much remained unresolved despite the huge number of applied research. Most modeling methods based on the concept of the diffusion of moisture through the material, describe a change in the average moisture content of the material over time [5].

The estimation of the effective diffusivity coefficient during the drying process of various food materials is very common in literature. Similar researches were conducted for potato material [1, 2, 4, 7, 9, 10]. The effect of air temperature, the thickness of slab and air velocity on the moisture diffusion was usually investigated.

However, unlike other, this chapter includes relative humidity besides all other drying parameters. The experiments were conducted in convective dryer with full air recirculation, which makes the control of the drying agent (air) relative humidity less complicated. This way, all important drying parameters were considered.

6.2   MATERIALS AND METHODS

Potatoes (var. *Désirée*) were used in the experiments. This variety of potatoes was introduced in Holland in 1962. It has great resistance to various diseases and it is suitable for growing in regions with low rainfall. The annual production in Republic of Serbia is 900,000 tons. It has excellent storage properties.

The usual shapes of the final product of dried potatoes are: french-fries shaped pieces and chips slices. Potato chips shape is chosen because of the simplicity and solvability of the diffusion model equations [3]. The side dimensions of 2 and 3 mm are chosen from the range between 0.3 and 4.3 mm, that is a normal range for this shape of the material widely used in literature [6, 7]. This geometry shape is suitable for describing in *Descartes* rectangular coordinate system. The material dimension in the direction of z-axis is much smaller than the other two dimensions in the direction of the x-axis and y-axis. Therefore, the main route of moisture transfer through the material will be in the direction of the z-axis. The moisture movement through the material in the direction of x-axis and y-axis may be omitted. This way the mass transfer equations will be simplified.

For samples preparation, potatoes were washed, peeled and cut into potato chips slices with 2 and 3 mm thickness. The chips slices were put in thin-layer on the 240 x 200 mm tray with net weight of 0.1 kg. The initial moisture content of apple cubes was determined using the oven-drying method [8] with repetition in order to assure accurate initial moisture content average values. The initial moisture content of the samples was found to be 76±0.3%, with initial moisture ratio of 3.58 kg water/kg dry matter.

6.2.1 EXPERIMENTAL APPARATUS

Convective laboratory scale *HPD* system with possibility of full drying air recirculation (Figure 6.1) was designed and built in order to achieve and measure all relevant drying process parameters, and with possibility of the full control of drying air RH. It consists of a closed loop tunnel sections with built-in drying chamber. Primary condenser (*CD1*) used for air heating is positioned at the top of the vertical tunnel section above the tray section (*TR*) that stands on digital weight indicator (*DWI*). Evaporator (*EV*) used for air cooling is positioned at the bottom of the horizontal tunnel section, separated from the by-pass (*BP*) line. Water condensed on

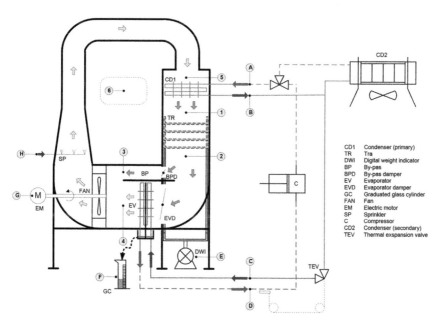

FIGURE 6.1 Laboratory HPD system with full air recirculation.

EV surface drains out of the system to the graduated glass cylinder (*GC*) located outside of the dryer. Desired drying air velocity is provided with an axial fan with electric motor frequency speed control unit. The amount of air that flows through *BP* and *EV* is regulated manually with by-pass damper (*BPD*) and evaporator damper (*EVD*).

The measurement equipment for the experiments has the following accuracies: temperature measurement: ±0.175°C, air relative humidity measurement: ±1%, drying air velocity measurement: ±0.01 m s⁻¹, weight measurement: ±0.001 kg, measurement: ±5 cm³, current frequency measurement: ±0.01 Hz. The following measurements were performed (Figure 6.1):

- at positions 1 to 5: drying air parameters (*T*, *w*, and *RH*);
- at position 6: ambient air properties (*T* and *RH*);
- at positions A, B, C and D: dryer components control parameters (temperatures at *CD1* and *EV* inlet/outlet connections);
- at position E: sample weight during the drying process;
- at position F: removed water volume;
- at position G: electric motor parameters (current frequency).

6.2.2 EXPERIMENTAL DESIGN

All measurements and experimental procedures were carried out in accordance with the methodology described by Zlatanovic [13]. Drying air temperature was in range 35–55°C, relative humidity in range 10–30% and velocity in range 1–2 m/s. Drying air was circulating over the wet material staggered on a tray.

6.2.3 DRYING MODELS

Fick's second law of unsteady state diffusion [Eq. (1)] for cube geometry, that provides connection between effective moisture diffusivity and moisture ratio [3], was used to interpret the drying process since moisture diffusion is one of the main mass transport mechanisms that describe this process.

$$\frac{\partial M}{\partial \tau} = D_{eff} \nabla^2 M = D_{eff} \left(\frac{\partial^2 M}{\partial x^2} + \frac{\partial^2 M}{\partial y^2} + \frac{\partial^2 M}{\partial z^2} \right) \quad (1)$$

In the case of symmetric boundary conditions, with neglecting of material shrinkage and with the assumption that water distribution in material (2 and 3 mm chips) is homogeneous, the moisture ratio can be determined [1–3] as in Eq. (2).

$$MR = \frac{M - M_e}{M_0 - M_e} = \frac{8}{\pi^2} \cdot \exp\left(-\frac{\pi^2 D_{eff} \tau}{4z^2}\right) \tag{2}$$

Taking into account the change of MR with drying time, Eq. (2) can be solved only numerically. In this case, a linear correlation between the natural logarithm of MR and time is obtained and can be used in Eq. (3).

$$\ln MR = n\frac{8}{\pi^2} + \left(-\frac{\pi^2 D_{eff}}{4z^2}\right) \cdot \tau \tag{3}$$

A plot of $\ln MR = f(\tau)$ gives a straight line with a slope that is used to determine [3] effective moisture diffusivity (D_{eff}) according to Eq. (4).

$$Slope = \frac{\pi^2 D_{eff}}{4z^2} \tag{4}$$

Drying rate of products during drying experiments was calculated by using the following equations:

$$DR = \frac{M_\tau - M_{\tau+\Delta\tau}}{\Delta\tau} \tag{5}$$

6.3 RESULTS AND DISCUSSION

The knowledge of MR change in time opens up the possibility of mutual comparison of the kinetics of drying of many different materials regardless of total weight of the material. Determination of the MR was performed on the basis of the experimental results and methodology [Eq. (2)] and presented on Figures 6.2 and 6.3. The value change of MR in the process follows the character of the changes of moisture content. The results of experimental research, indicate the lack of a constant-rate period of drying, i.e., only the existence of the period of falling rate of drying (Figures 6.4 and 6.5).

During the drying process, the drying rate is declining in a constant, which can be explained by the fact that the material is dried in a thin layer. The highest drying rate occurs at the beginning of drying, when

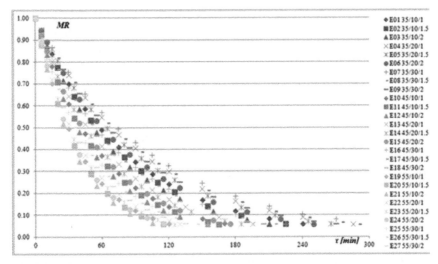

FIGURE 6.2   Effect of temperature, relative humidity and velocity of drying air on MR change in time (Potato, chips 2 mm).

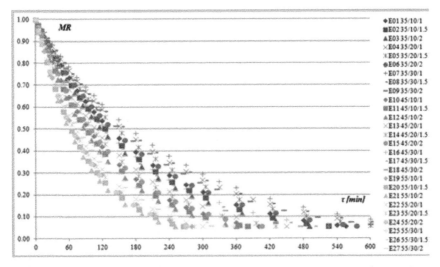

FIGURE 6.3   Effect of temperature, relative humidity and velocity of drying air on MR change in time (Potato, chips 3 mm).

the average moisture content is highest. This is explained by the position of the zone of vaporization, which is closer to the outer surface of the material at the beginning of the process, so the resistance to the transfer of moisture through the material is smaller.

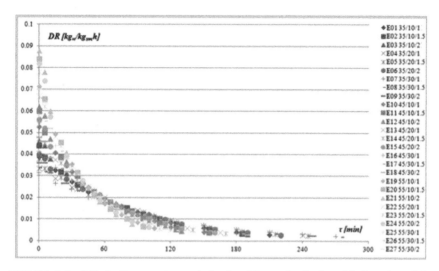

FIGURE 6.4   Effect of temperature, relative humidity and velocity of drying air on DR change in time (Potato, chips 2 mm).

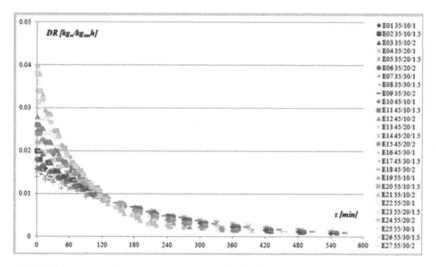

FIGURE 6.5   Effect of temperature, relative humidity and velocity of drying air on DR change in time (Potato, chips 3 mm).

By using the effective diffusion coefficient it is possible to include all the possible mechanisms of the transfer of moisture that occur in the material, so that the transfer of moisture depends only on the geometrical shape and dimensions of materials which influence the direction and the direction of movement of moisture. In the case of material of potato, sliced into thin sheets (chips), the thickness of the sheets (in the z direction) is several times smaller than the other two dimensions (in directions x and y-axis), so that the direction of movement of moisture takes the direction of the z-axis. The movement of moisture in the directions of the x and y-axis is ignored. The characteristic length of material z that exists in the Eq. (2) in this case represents the ½ of the chips thickness. Determination of the effective diffusion coefficient in this case is based on the assumption that the transmission of moisture takes place in the two directions along the direction of the z-axis. The calculated values of effective diffusivity coefficient for each experimental setup are presented in Table 6.1.

The values for the potato effective diffusivity coefficient are in range: $D_{eff} = 1.50 \times 10^{-8} \div 4.28 \times 10^{-8}$ [m²/s]. The effective diffusion coefficient increases with increasing temperature humidity and velocity of moist air, and decreases with increasing relative humidity and the characteristic dimensions of a piece of material. In Table 6.2, the effective diffusion coefficient are presented in a such a way that all drying parameters that have negative influence (relative humidity and material size) on D_{eff} are in columns and all drying parameters with positive influence (temperature and velocity) on D_{eff} are in rows.

The highest values of effective diffusion coefficients were determined in experiments E21 with the set parameters of drying: $T = 55°C$, $RH = 10\%$, $w = 2$ m/s. This is explained by the maximum values of the parameters of temperature and flow velocity of moist air, and minimum of the relative humidity value, in the observed range of size variation. The lowest values of effective diffusion coefficients were determined in experiments E7 with the set parameters of drying: $T = 35°C$, $RH = 30\%$, $w = 1$ m/s. This is explained by the minimal values of the parameters of temperature and flow velocity of moist air, and the maximum value of the parameter relative humidity, in the observed range of size variation. The influence of drying parameters variation on the change of effective diffusivity coefficient is presented in Table 6.3.

Drying time was measured for potato sample dimensions of 2 mm and 3 mm in different experimental setups (Figure 6.6). The maximum drying time was observed in the experiment E7 and the minimal in E21.

TABLE 6.1 The Results for Effective Diffusivity Coefficient for Each Experimental Setup

| Experimental setup | Drying regime | | | | |
	T [°C]	RH [%]	w [m/s]	z =2 [mm]	z =3 [mm]
E1	35	10	1	1.91	1.90
E2	35	10	1.5	2.04	2.01
E3	35	10	2	2.17	2.12
E4	35	20	1	1.64	1.61
E5	35	20	1.5	1.75	1.72
E6	35	20	2	1.87	1.83
E7	35	30	1	1.51	1.50
E8	35	30	1.5	1.62	1.57
E9	35	30	2	1.72	1.68
E10	45	10	1	2.68	2.63
E11	45	10	1.5	2.82	2.77
E12	45	10	2	2.97	2.92
E13	45	20	1	2.39	2.34
E14	45	20	1.5	2.50	2.45
E15	45	20	2	2.63	2.59
E16	45	30	1	2.04	2.01
E17	45	30	1.5	2.14	2.12
E18	45	30	2	2.26	2.23
E19	55	10	1	3.83	3.72
E20	55	10	1.5	4.04	3.94
E21	55	10	2	4.28	4.20
E22	55	20	1	3.54	3.47
E23	55	20	1.5	3.73	3.65
E24	55	20	2	3.98	3.87
E25	55	30	1	3.00	2.96
E26	55	30	1.5	3.18	3.10
E27	55	30	2	3.38	3.32

TABLE 6.2   Effective Diffusivity Coefficient Deff $\times 108[m^2/s]$

Drying Parameters		z=2 mm			z=3 mm		
		RH=20	RH=30	RH=10	RH=10	RH=20	RH=10
m/s	°C	%					
w = 1	T=35°C	1.91	1.64	min. 1.51	1.90	1.61	min. 1.50
	T=45°C	2.68	2.39	2.04	2.63	2.34	2.01
	T=55°C	3.83	3.54	3.00	3.72	3.47	2.96
w = 1.5	T=35°C	2.04	1.75	1.62	2.01	1.72	1.57
	T=45°C	2.82	2.50	2.14	2.77	2.45	2.12
	T=55°C	4.04	3.73	3.18	3.94	3.65	3.10
w = 2	T=35°C	2.17	1.87	1.72	2.12	1.83	1.68
	T=45°C	2.97	2.63	2.26	2.92	2.59	2.23
	T=55°C	max. 4.28	3.98	3.38	max. 4.20	3.87	3.32

TABLE 6.3   The Influence of Drying Parameters Variation on Effective Diffusivity Coefficient

Drying parameter	The variation of drying parameter	Potato chips, 2 mm
Temperature	+60%	+100%
Relative humidity	+100%	−14%
Air velocity	+100%	+13%

6.4   CONCLUSIONS

In this chapter, the effects of drying agent (air) temperature, relative humidity and velocity as well as the material dimension influence on the kinetics of the drying process was investigated. Drying air temperature was in range 35–55°C, relative humidity in range 10–30% and velocity in range 1–2 *m/s*. Potatoes were sliced into chips with the thickness of 2 and 3 mm.

It was found that regardless of the dimension of the test material, the drying regime, with the highest temperature and velocity agents, and minimum relative humidity, achieves the shortest possible duration of the drying process, and vice versa.

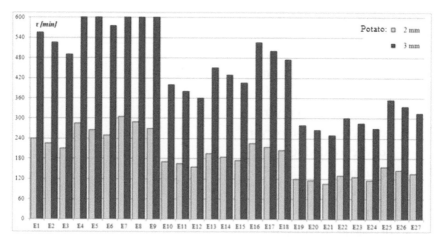

FIGURE 6.6 Drying Time for Each Experiment (Potato chips 2 and 3 mm).

The influence of individual parameters on the kinetics of drying scratched discussed over the obtained values of effective diffusion coefficient. The effective diffusion coefficient does not change in time. The increase of the air velocity and temperature agent leads to a rise in effective diffusivity coefficient values with predictable cause-and-effect relationship. Obtained values of the effective diffusivity coefficient are in the range: $1.50 \times 10^{-8} \leq D_{eff} \leq 4.28 \times 10^{-8} \ m^2/s$.

6.5 SUMMARY

An experimental investigation of convective drying of *Désirée* potatoes in full recirculation laboratory scale dryer was conducted. Potatoes sliced into chips (thickness 2 and 3 mm), without pre-treatment, were used. The parameters of air stream over the tray with samples were fully controlled and adjusted in several sets of experiments. Drying air temperature was in range 35–55°C, relative humidity in range 10–30% and velocity in range 1–2 *m/s*. Data were analyzed to obtain diffusivity values from the period of falling drying rate. The effective moisture diffusivity was used to describe drying process efficiency. The obtained values for the potato effective diffusivity coefficient were in range $1.50 \times 10^{-8} \div 4.28 \times 10^{-8}$ [m^2/s].

KEYWORDS

- agricultural engineering
- apple
- continuous band dryer
- convective cyclone dryer
- convective drying
- diffusion
- diffusivity
- drying
- drying air parameters
- drying kinetics
- effective moisture diffusivity
- experimental results
- Fick's Second Law
- food processing
- heat and mass transfer
- potato
- potato chips
- potato modeling
- Rusia
- shrinkage
- single layer drying
- sweet potato
- thin-layer drying
- tomato

REFERENCES

1. Aghbashlo, M., Kianmehr, M. H., & Arabhosseini. A. (2009). Modeling of thin-layer drying of potato slices in length of continuous band dryer. Energy Conversion and Management, 50, 1348–1355.
2. Akpinar, E., Midilli, A., & Bicer Y. (2003). Single layer drying behavior of potato slices in a convective cyclone dryer and mathematical modeling. Energy Conversion and Management, 44, 1689–1705.

3. Crank, J. (1975). The Mathematics of Diffusion. Second edition, Brunel University Uxbridge, Oxford University Press, London. ISBN 0-19-853344-6.
4. Hassini, L., Azzouz, S., Peczalski, R., & Belghith, A. (2007). Estimation of potato moisture diffusivity from convective drying kinetics with correction for shrinkage. Journal of Food Engineering, 79(1), 47–56.
5. Kemp, I., Fyhr, B., Laurent, S., Roques, M., Groenewold, C., Tsotsas, E., Sereno, A., Bonazzi, C., Bimbenet, J., & Kind, M. (2001). Methods for processing experimental drying kinetics data. Drying Technology, 19(1), 15–20.
6. Kingsly, A. R. P., Singh, R., Goyal, R. K., & Singh, D. B. (2007). Thin-layer drying behavior of organically produced tomato. American Journal of Food Technology, 2(2), 71–78.
7. Leeratanarak, N., Devahastin, S., & Chiewchan, N. (2006). Drying kinetics and quality of potato chips undergoing different drying techniques. Journal of Food Engineering, 77, 635–643.
8. Lengyel, A. (2007). The change of body temperature during convective drying of cube-shaped apple pieces. Drying Technology, 25(7–8), 1275–1280.
9. Simal, S., Rossello, C., Berna, A., & Mulet, A. (1994). Heat and mass transfer model for potato drying. Chemical Engineering Science, 49(22), 3739–3744.
10. Singh, J. N., & Pandey, R. K. (2011). Convective air drying characteristics of sweet potato cube (Ipomoea batatas L.). Food and Bioproducts Processing, doi: 10.1016/j. fbp.2011.06.006.
11. Zlatanović, I. (2012). Application of modern drying technology in the food processing industry. Scientific Journal Agricultural engineering, 37(4), 23–30.
12. Zlatanović, I. (2012). Types, classification and selection of dryers in agro industry. Scientific Journal Agricultural Engineering, 37(2), 1–13.
13. Zlatanovic, I., Komatina, M., & Antonijevic, D. (2013). Low-temperature convective drying of apple cubes. Applied Thermal Engineering, 53(1), 114–123.

PART 3

WATER QUALITY AND MANAGEMENT

CHAPTER 7

SUB-SURFACE DRAINAGE SYSTEMS AND FLOW MONITORING IN CANALS FOR INCREASING CROP PRODUCTIVITY IN AGRICULTURE

K. PALANISAMI, R. DORAISWAMY, and M. K. KHAISHAGI

CONTENTS

7.1 INTRODUCTION

Excessive and uncontrolled use of irrigation in water intensive crops like paddy and sugarcane, clogging and siltation in natural drains and nalas, seepage from canal network, presence of soils having low permeability, low lying areas in the vicinity of rivers and streams having shallow water table and poor irrigation practices are among the factors causing water logging in the irrigation canal project areas [1, 2, 4]. For example, an increasing extent of area in the Narayanpur Right Bank Canal and the

Ghataprabha Project commands is facing water logging problem, leading to low fertility and large segment of agricultural land turned out of cultivation. As the production decreases incrementally over the years, farmers suffer economically thereby affecting the overall economic growth and food security of the locality [3].

The soils of Ghataprabha area are mostly Deccan trap, medium to deep calcareous, clay in texture, have poor drainage characteristics. Small areas of alluvial soils are found along river beds. At several places B.C. soil is present, including Jamkhanditehsil that is prone to water logging. About 22,000 hectares area in Ghataprabha project is affected by water logging. Nayayana Right Bank Canal has 3,627 ha water logged area in Deodurg and Lingsugur, whereas 22 ha has been reclaimed so far (Table 7.1). About 32,567 hectares command area in Ghataprabha project is affected by water logging, salinity and alkalinity (Table 7.2). An area of 676.94 ha is reclaimed so far since inception. Affected areas in these two projects are scattered in patches at various vulnerable places.

7.2 SUB-SURFACE DRAINAGE

Drainage is the natural or artificial removal of surface and sub-surface water from an area. Many agricultural soils need drainage to improve production or to manage water supplies. There are two main methods available for land reclamation namely: surface drainage and subsurface drainage [5–7].

7.2.1 SUBSURFACE DRAINAGE SYSTEM (SSD)

This is an effective engineering intervention to drain off subsurface water, using perforated piped network buried in the soil at about 1.2 m below ground level. A Lateral interval of 21 m is found tobe satisfactory. Laterals are connected to the sub-mains, which in turn drain into the mains. Finally main pipelines discharge into suitably deepened natural drains and *nala*. General slope of the subsurface piped network is in the direction of natural ground slope, with laterals mostly perpendicular to contour

TABLE 7.1   Water Logged, Salinity and Alkalinity Status in NRBC (ha)

Parameter	Area, ha	
	Deodurg	Lingsugur
Alkaline	2381	255
Area reclaimed from Salinity	-	47
Area reclaimed from water logging	1078	377
Area reclaimed from Alkalinity	-	22
Saline	1939	402
Water logged	1964	1663

TABLE 7.2   Water Logged, Saline and Alkaline Affected Area in Ghataprabha Project

Tehsil	Existing affected area in ha (since inception)			
	Water logged	Saline	Alkaline	Total
Athani	2350	1500	-	3850
Bilagi	2620	1460	-	4080
Chikkodi	350	450	-	800
Gokak	4580	2670	-	7250
Hukkeri	911	570	-	1481
Jamkhandi	1989	2587	-	4576
Mudhol	1710	3025	595	5330
Raibag	2970	2230	-	5200
Total	**17,480**	**14,492**	**595**	**32,567**

Note: Tehsil – It is an area of land with a town that serves as its administrative center, with possible additional towns, and usually a number of villages. The terms in India have replaced earlier geographical terms, such as *pargana, pergunnah* and *thannah*, used under the Delhi Sultanate and the British Rule in India.

lines. As an example, a significant part of water logged area is reclaimed in Jamakhanditehsil under Ghataphrabha Left Bank Canal Command Area, and the results are satisfactory. Figure 7.1 presents classification of drainage systems in agriculture. Figure 7.2 presents selected examples of the positive impact of land reclamation programs.

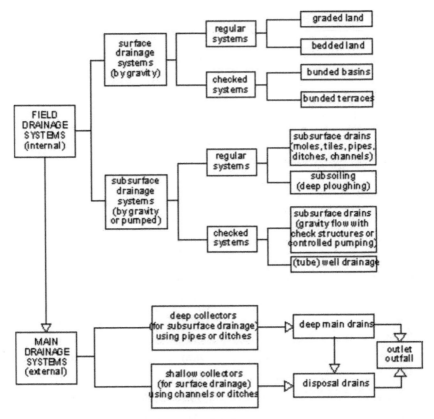

FIGURE 7.1   Classification of agricultural drainage systems.

7.2.2 ECONOMICS OF SUB-SURFACE DRAINAGE SYSTEMS

Financial analysis of subsurface drainage systems indicates internal rate of return (IRR) as 32% and benefit cost ratio (BCR) of 2.98 for soybean crop assuming 10 years lifespan for SSD. Corresponding values for sugarcane are much higher as 61% IRR and 5.30 BCR, which are expected. Cost estimates indicate that laying of pipes for SSD system costs 48% of the total cost, while supply of corrugated perforated and non-perforated pipes and fittings accounts for 28%. Providing and fixing of enveloping material costs 19%, construction of structures for SSD system 3% and miscellaneous expenditure 2%. Most estimates indicate that SSD costs near about forty thousand rupees per hectare (60.00 Rs. = 1.00 US$).

Drained off water from subsurface drainage system as
it comes out from submains, Jamkhandi Tahala

Soyabean crop in reclaimed water logged area
after subsurface drainage, Jamkhandi Tahala

Adjacent farms show contrast between reclaimed water logged
area and drained land, Jamkhandi Tahala

Drained off water from syb surface drainage system
flowing in open channel, Jamkhandi Tahala

Untreated water logged area shown as wasteland,
Jamkhandi Tahala

Collection well for subsurface drainage system
reuse of drainage water, Jamkhandi Tahala

Subsurface drainage water reused by
disel pump set, Jamkhandi Tahala

FIGURE 7.2 Selected examples of the positive impact of land reclamation.

7.2.3   GOVERNMENT SUBSIDY ON SSD

Reclamation of water-logged area in the irrigation command has been included in the centrally sponsored command area development programs:

- to reclaim irrigation land affected due to water logging, salinity and alkalinity.
- reclamation is done either by surface drains or by sub surface drain.
- the assumed cost of surface drainage system is Rs. 15,000/ha and for sub surface it is Rs. 40,000/ha.
- the extent of central assistance is Rs. 7,500/ ha or 50% of actual expenditure whichever is less in case of surface drainage and Rs. 20,000/ha or 50% of actual expenditure whichever is less in case of sub surface drainage.
- prior approval of GOI for the proposal is necessary before taking up the works.
- a minimum of 10% contribution (of the total cost) by the beneficiary farmer as a part of the State share is mandatory.

7.3   SURFACE DRAINAGE SYSTEM

Several affected areas in Ghataprabha Project command need surface drainage system. This system involves construction of open drains in the affected area that can safely remove the excess water into the natural drains and nala. Surface drainage system is simpler and cheaper compared to sub-surface drainage however it consumes more land area that goes into construction of open drains. The drawings related to surface and subsurface drainage are given in Figures 7.3 and 7.4.

During the discussions with the stakeholders, it was observed that about a quarter of the water logged area can be reclaimed using surface drainage system in Ghataprabha Project. Analysis of the cost estimates indicate that over 80% expenditure is for excavation for construction of drains, 17% towards construction of structures like culverts and reinforced cement concrete hume pipes, etc. Components like bushes and jungle clearing and removal of girth and tree cutting, etc. involves other miscellaneous expenditure. The details of the cost and benefits of surface and sub-surface drainage are given in Tables 7.3–7.5. Total cost towards providing drainage of water-logged area is Rs. 75.47×10^7.

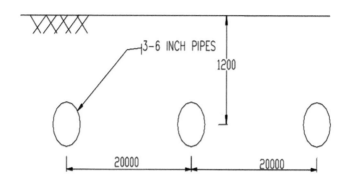

GROUND LEVEL

3-6 INCH PIPES

1200

20000 20000

SUB-SURFACE DRAINAGE SYSTEM

DRAINAGE PARTICULARS:
LATERAL INTERVAL: 20000 MM
DEPTH BELOW GL: 1200 MM
DIA OF PIPES: 75-150 MM

FIGURE 7.3   Typical layout of subsurface drainage system.

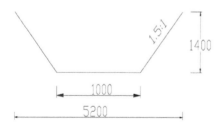

1.5:1

1400

1000

5200

SURFACE DRAIN SECTION

DRAIN PARTICULARS:
BASE WIDTH: 1000 MM
DEPTH: 1400 MM
TOP WIDTH: 5200 MM
SIDE SLOPE: 1.5:1

FIGURE 7.4   Typical layout of surface drainage system.

TABLE 7.3   Cost of Surface and Sub-Surface Drainage, Rs. per ha (60.00 Rs. = 1.00 US$)

Surface drainage (SD)	
Area covered under surface drains = top width x length of drain (sq.m)	5.20 m x 4500 m
Value of agricultural land (Rs/ha)	1,250,000
Land value in fields drains (Rs)	2,925,000
Total land area (ha)	193.49
Cost of land used in drains (Rs/ha)	15,117.06
Gross cost of land reclamation (Rs/ha)	15,117.06 + 8000
Value of the area foregone due to drains (Rs./ha)	23,117.06
Sub Surface Drainage (SSD)	
Typical cost (Rs./ha)	40,000
Life span (years)	10.00
Soybeanincome before SSD, (Rs./ha)	18,750
Soybeanincome after SSD, (Rs./ha)	65,000
Net benefit (Soybean), (Rs./ha)	46,250
Sugarcane yield before SSD (t/ha)	62.5
Sugarcane yield after SSD (t/ha)	95
Sugarcane price, (Rs./ton)	2,000
Benefit of SSD (Rs./ha)	65,000

7.3.1   IMPLEMENTATION STRATEGY

7.3.1.1   Public Private Partnership

In view of the high cost of pipes and fittings, it appears feasible to develop *Public Private Partnership* (PPP) between pipe manufacturers having marketing opportunities, and sugar mills which are beneficiaries of the sugarcane production on the one hand and farmers on the other. High IRR

TABLE 7.4 Sub-Surface Drainage (SSD)

S. No.	Particular	Ghataprabha	NRBC
1	Total water logged area ha	17,480	3,627
2	Proposed subsurface drainage area ha	13,110	2,720
3	1st Year SSD, ha	7,000	1,500
4	2nd Year SSD ha	6,110	1,220
5	Unit rate Rs./ha	40,000	40,000
6	Total SSD cost Rs.	52.44×10^7	10.88×10^7
7	IRR:		
	a) Soybean	24%	
	b) Sugarcane	39%	
8	B/C Ratio		
	a) Soybean	2.43	
	b) Sugarcane	3.45	

TABLE 7.5 Surface Drainage

S. No.	Particular	Ghataprabha	NRBC
1	Total water logged area ha	17,480	3,627
2	Proposed area of surface drainage (SD) ha	4,379	907
3	1st Year SD, ha	2,300	500
4	2nd Year SD, ha	2,079	407
5	Unit rate Rs./ha.	23000	23000
6	Total SD cost Rs.	10.07 Crores	2.08 Crores
7	IRR:		
	a) Soybean	22%	
	b) Sugarcane	33%	
8	B/C Ratio:		
	a) Soybean	2.67	
	b) Sugarcane	3.83	
9	Total Cost = SD + SSD, Rs.	62.51×10^7	12.96×10^7

and BCR offer reasonably safe investment and business opportunities for these players.

7.3.1.2   Farmers Participation

The proposal may preferably be tried in areas where Water Users Cooperative Societies are well developed and active thereby giving better opportunities to involve farmers in identifying areas, planning, implementing and maintenance of SSD systems. WUCS are well placed to take-up labor intensive activities like excavation of drains and clearing of jungle, bushes and desilting activities.

7.4   CANAL FLOW MEASUREMENT

7.4.1   MAIN CANAL IRRIGATION SYSTEM

Our impressions regarding the Ghataprabha Project infrastructure during field visit are positive. The main canal in general is in good condition and the present proposal for lining and remodeling works appear to be a good investment in the durability and reliability of the system. Scope for further improvement of the main canal is limited, but secondary canals (distributaries, minors) can be further improved to take full advantage of the system. However, operational practices have not kept with the improvements in the system and opportunities have not been fully utilized. Therefore, this also asks for more real-time monitoring of irrigation flows and more modern operation and maintenance (O&M), giving quicker response to the needs of farmers. The memorandum of understanding between WRD and WUCs contains a clause about volumetric water supply.

The present system of water level recording of main canal system does not give the required feed-back for dynamic O&M. Automatic flow measurement devices are recommended at strategic places in the main canal, plus a system of automatic collection and processing of this information. This will enable real-time monitoring and control. Improved water use efficiency needs flow measurement as the basis of O&M in a modern, lined irrigation system. There is big scope for further improvement of the distributaries and minors (secondary system) in creating conditions for volumetric water management instead of maintaining water levels. Investments are possible in detailed improvement of drops and weirs and regulating structures to allow volumetric measurement. Automatic (SCADA) systems may be added at strategic points in secondary canals as well.

7.4.2 VOLUMETRIC BASED SUPPLY

The ultimate aim of the proposed intervention is that each water user has to reliably know the quantities of water received and it should be expressed in such a way that is comprehensible and meaningful to them. It was noted during the field visit that decades old concept of volumetric based irrigation water management is neither fully internalized by the field staff nor by the farmers. The present system is based on ayacut area and crop type, thus encouraging the farmers to consume as much water as possible. The rules of the game must be modified so that judicious use of water will be more rewarding, and wastage will be just that. Presently neither engineers nor farmers have any awareness about quantities of water supplied and consumed. In GLBC command area, 6 Reinforced Fiber Plastic Cut Throat Flumes are installed on experimental basis to measure flow to WUCS and these are working satisfactorily. Figure 7.5 shows one such flume we visited during our field trip.

7.4.3 FLOW MEASUREMENT METHODS

Canal flow measurement is possible by several methods like measuring flumes, current meter, stage discharge method, notch and weirs, hydraulic structures like canal drops and electronic equipments like SCADA. It has been noticed that measuring devices are mostly installed at the head of the distributaries and in several cases, gauge registers are maintained and readings are recorded round the clock. Measuring flumes at strategic points along big distributaries, minors, and also at the interface of WUCs may be considered for installation. Currently, the WUCs do not have control of the gates of the canal nor do they have measuring devices to monitor water discharge. Monitoring by KNNL officials could check the overuse

Reinforced fiber plastic cut throat flume on OT of
Budhni minor canal: GLBC

Standing wave flume along the Ghataprabha left bank canal.
GLBC-standing wave flume and road bridge

GLBC -guage well for standing wave flume

GLBC - stone masonry standing wave flume and
gauge well on minor canal

FIGURE 7.5   Flow measuring devices.

of water but there is no incentive for these officials to do so. In fact, few officials claimed that they have faced life threat from farmers for checking water theft from main canal during night time. In short, the supply based on the original plan falls short of the current demands. The following two approaches are being proposed for this project:

a. Flow measuring flumes

Along the main canal and distributaries, standing wave flumes are constructed and these are mostly in good working condition. Gauge reading are regularly taken and recorded in gauge registers during irrigation season. However, the data is not properly digitized and used for analysis, planning and operation.

Flow measurement is first step towards increasing productivity of available water resource to produce more food with same quantity of water. Increased water productivity can provide means to ease water scarcity in tail and middle reaches. Serving growing population and competing sectors of water user (Drinking, Irrigation, power, Industries and other users) needs conservation and optimal utilization of available water resources.

b. Doppler based online flow monitoring

Several developments in recent years have made SCADA based real-time monitoring and control cost-effective for even smaller irrigation systems. The convergence of the following technologies has made low-cost automation a viable reality: (1) low-cost data loggers/controllers; (2) a growing variety of inexpensive sensors; (3) expanding use of solar-energy systems; (4) innovations in communication equipment; (5) rapid advancements in the PC industry; and (6) the phenomenal growth of the Internet.

c. GLBC: Online Doppler based flow monitoring system

A Doppler based flow monitoring system is installed in GLBC canal network having first measuring point at Hidkal Dam monitoring real – time reservoir levels and capacity. This system covers 50 strategic monitoring points along the main canal and distributaries. The total command area of GLBC is 1,61,887 ha covered under 150 distributaries. However, 75% of water in is conveyed through 45 large distributaries. Hence this system is able to continuously monitor 75% of flow in command area along with 5 major milestones on the main canal. The sensor units, based on side looking Doppler, are powered by solar panels installed nearly. Flow monitoring data is transmitted through GSM/GPRS network to central server located at KNNL Office in Bangalore. Processed information is communicated to concerned officials and water users in the form of SMS as well as online (Figure 7.6). Water loss is minimized through precession control and detection.

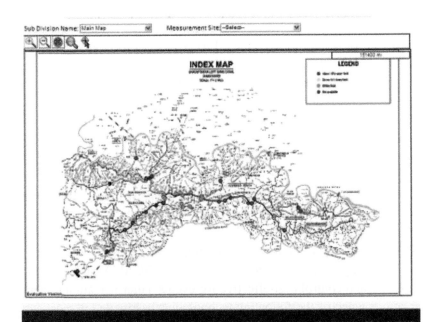

GLBC doppler based online flow monitoring system

Damaged canal network
Senor unit of doppler based online flow monitoring system, also seen id damaged canal section: GLBC

FIGURE 7.6   GLBC: Online Doppler based flow monitoring system.

7.4.4 THE OPERATION OF THE SYSTEM IN BRIEF

The system as a whole helps capture the water requirement in discharge terms, across the distributaries, branches and reaches of the network. Once the water requirement or plan is captured into the software component of the system, the flow monitoring field station sensors will monitor the flow accurately and communicate the same in real time to a central monitoring station. The real time data is compared with the captured plan to assess the deviation in flow as against the plan. The deviations are then communicated to the relevant officers and operators for further actions to bring the flow to the limits as per the plan. The deviations are escalated to the superior officers if there are repeated violations. The details of the cost of the structures are given in Table 7.6.

7.4.5 BENEFITS OF FLOW MEASUREMENT

In the existing online flow monitoring system 15% saving in water usage is assumed, however our assessment indicate about 10% savings, and thus water savings through this proposal works out to 7.78 TMC per annum. Besides the proposal gives more confidence to WUCs to adopt volumetric

TABLE 7.6 Summary of the Flow Measurement Proposal

S. No.	1st Year		2nd Year		3rd Year		Total Nos.	Total Cost, Rs.
	Nos.	Cost, Rs.	Nos.	Cost, Rs.	Nos.	Cost, Rs.		
GLBC Online	30	15,000,000	30	15,000,000	40	20,000,000	100	50,000,000
GLBC+ GRBC CT Flumes	50	1,250,000	100	2,500,000	97	2,425,000	247	6,175,000
GRBC Online	40	1,000,000	Nil	Nil	Nil	Nil	40	40,000,000
NRBC CT Flumes	50	1,250,000	67	1,675,000	70	1,750,000	187	4,675,000
Total	170	18,500,000	197	19,175,000	207	24,175,000	574	100,850,000

supply and help in achieving more equitable and judicious water usage. Basic objective is to deliver adequate quantity of water on time to all users equitably within the command area. Monitoring of flow continuously, concurrently and accurately helps the management to optimally deliver water to all users at right time, and help to vary the flow according to need. Continuous monitoring keeps track of any deficiency in supply and bring it back to the planned level for compliance.

7.4.6   IMPLEMENTATION

Doppler based Online Flow Measurement as well as RFP Flumes shall be installed, maintained and operated by the WRD and KNNL engineers, in consultation with the WUCs, since they form interface between the two. Local suppliers are available who can also do the installation and mainte-nance for 5 years. The quantities supplied to the WUCs as well as timing shall be known and impose accountability and responsibility on all sides. In due course O&M may be transferred to WUCs.

7.5   SUMMARY

This chapter presents state of art for sub-surface and surface drainage systems and flow monitoring in canals for increasing crop productivity in agriculture. Two types of drainage systems are presented. The chapter also includes status of areas under water logging, salinity and alkalinity; economics of both types of drainage systems. Flow measuring devices includes flumes, Doppler based online flow measurement and its benefits.

KEYWORDS

- alkalinity
- benefit cost ratio
- canal network
- command area

- crop productivity
- Deccan trap
- drainage
- fertility
- flumes
- food security
- irrigation practices
- paddy
- permeability
- poor drainage
- salinity
- seepage
- soybean
- sub surface drainage
- sugarcane
- surface drainage
- water logging
- water table

REFERENCES

1. De Jong, M. H. (1979). Drainage of structured clay soils. In: Proceedings of the International Drainage Workshop. ILRI Publication 25, J. Wesseling (ed.), pp. 268–280. International Institute for Land Reclamation and Improvement (ILRI), Wageningen, The Netherlands.
2. Oosterbaan, R. J., Gunneweg, H. A., & Huizing, A. (1987). Water control for rice cultivation in small valleys of West Africa. In: ILRI Annual Report 1986, pp. 30–49. International Institute for Land Reclamation and Improvement (ILRI), Wageningen, The Netherlands.
3. Palanisami, K. (2012). Water resources group: expert input on preparing detailed project report for water-enabled growth in Karnataka. Water Management Interventions for Narayanpur Right Bank Canal and Ghataprabha Project in Krishna Basin in Karnataka. Unpublished report. IWMI, Hyderabad.
4. Qorani, M., Abdel Dayem, M. S., & Oosterbaan, R. J. (1990). Evaluation of restricted subsurface drainage in rice fields. Symposium on land drainage for salinity control

in arid and semi arid regions, volume 3. Drainage Research Institute, Cairo, pp. 415–423.

5. Rao, K. V. G. K., Sharma, D. P., & Oosterbaan, R. J. (1992). Sub-irrigation by groundwater management with controlled subsurface drainage in semi-arid areas. International Conference on Supplementary Irrigation and Drought Management, Bari, Italy.

6. Ritzema, H. P. (1994). Drainage Principles and Applications. ILRI Publication 16, pages 635–690. International Institute for Land Reclamation and Improvement (ILRI), Wageningen, The Netherlands.

7. Sevenhuijsen, R. J. (1994). Surface drainage systems. In: Drainage Principles and Applications, ILRI Publication 16, H. P. Ritzema (ed.), pp. 799–826. International Institute for Land Reclamation and Improvement (ILRI), Wageningen, The Netherlands.

CHAPTER 8

WATER QUALITY OF SHALLOW GROUNDWATER: CASE STUDY

MAYANK KUMAR and NITIN DABRAL

CONTENTS

8.1 INTRODUCTION

Water is one of the most important components of our environmental resources, for all living organisms. Survival of human and socio-economic development depends largely on the availability and use of water in agriculture industry. The evolution of irrigated agriculture in large alluvial basin traces the interaction of human society with nature. To begin with, water was needed largely for irrigating crops; public health, industry and recreation claimed a very small percentage of the total water supply. The water needs increased with time to meet the requirement of more food and industries with increasing population. Ground water (GW) resources meet the increasing demand of water for domestic and industrial purposes. Water pollution is a phenomenon characterized by deterioration of the quality as a result of various human activities. The poor quality of drinking water in our country is more due to contamination than due to inferiority of the sources. It has been estimated that about 25% of the irrigated land of world is affected to some degree by water salinity.

There are various parameters which determine water quality for various uses. These are: odor, color, taste, acidity, alkalinity, hardness, total dissolved solids, etc. These parameters should have certain normal value. If these values deviate from their normal values, the quality of water changes (i.e., change of water quality negatively is synonymous with degradation and pollution of water).

Water pollution occurs when waste products or other substances change the chemical or biological characteristics of the water and degrade water quality so that animals, plants or human uses of the water are affected. Pollutants include plant nutrients, bacteria, viruses, pesticides, herbicides, hydrocarbons (including petrol and oil), heavy metals and other toxic chemicals. Shallow groundwater is often affected by land use. Chemicals or microorganisms (bacteria and viruses) may filter through the soil to shallow water table through sandy soils. Groundwater in deeper (confined) aquifers beneath layers of rock or clay that do not let water through has better protection from pollution because it is not directly connected to the surface environment.

Agriculture, including commercial livestock and poultry farming, is the source of many organic and inorganic pollutants in surface waters and groundwater. These contaminants include both sediment from the erosion of cropland and compounds of phosphorous and nitrogen that partly originate in animal wastes and commercial fertilizers. Animal wastes are high in oxygen demanding material, nitrogen and phosphorous and often harbor pathogenic organisms. Wastes from commercial feeders are contained and disposed of on land; their main threat to natural waters, therefore, is via runoff and leaching.

Contamination can enter the water bodies through one or more of the following ways.

1. Diffuse agricultural sources: wash off and soil erosion from agricultural lands carrying material applied agricultural use, mainly fertilizers, herbicides and pesticides.

2. Diffuse urban sources: run off from city streets, from horticultural, gardening and commercial activities in the environment and from industrial sites and storage areas.
3. Direct point source: transfer of pollution from municipal industrial liquid waste disposal sites and from municipal and household hazardous waste and refuse disposal sites.

The processes of bioaccumulation and bio magnifications are extremely important in the distribution of toxic substances (discharged in waste effluents) in fresh water ecosystems. The concentration of pollutants within the organism due to bioaccumulation and bio magnifications depends on the duration of exposure of the organism to the contaminated environment and its tropic level in the food chain. Several fold increases in trace contaminants have been commonly observed in lakes and estuarine environments.

The poor quality of drinking water in our country is more due to contamination than due to inferiority of the sources. Agriculture, including commercial livestock and poultry Farming, is the source of many organic and inorganic pollutants in surface waters and groundwater. These contaminants include both sediment from the erosion of cropland and compounds of phosphorous and nitrogen that partly originate in animal wastes and commercial fertilizers. Animal wastes are high in oxygen demanding material, nitrogen and phosphorous and often harbor pathogenic organisms. Wastes from commercial feeders are contained and disposed of on land; their main threat to natural waters, therefore, is via 'runoff and leaching. Control may involve settling basins fro liquids, limited biological treatment in aerobic or anaerobic lagoons and a variety of other methods.

The degradation of water quality has severe effects in context with drinking, agriculture, and industrial purposes. Drinking impure water results in various harmful diseases like diarrhea, blue baby disease in case of excess of nitrite in water. In industries if water has more hardness due to chloride and sulfide, it is not suitable for cleaning purposes. In agriculture, if water has more salt concentration then it reduces permeability of soil and infiltration is reduced. More Nitrate concentration is harmful for dirking but safe for irrigation. There are several other effects, which should be looked into.

The present study was conducted in the Sitarganj area of Udham Singh Nagar with the following objectives:

1. To study the physico-chemical characteristics of water from the shallow aquifer,
2. To study the extent of pollution and to judge the suitability of water for irrigation, industrial and domestic use.

8.2   REVIEW OF LITERATURE

The rapid pace of urbanization/industrialization as well as agricultural activities have made environmental pollution a growing concern globally. Off all the receptor system exposed to the contaminants, ground water has received little attention in the past because of common belief that ground water was pristine.

Ground water pollution is usually traced back to four main origins industrial, agricultural, domestic, and over exploitation. Studies carried out in India reveal that one of the most important causes of ground water pollution is unplanned urban development without adequate attention to sewage and waste disposal. Industrialization without provision of proper treatment and disposals wastes and effluent is another source of ground water pollution. Excessive applications of fertilizers for agricultural development coupled with over-irrigation intrusion due to excessive pumping of fresh water in coastal aquifers are also responsible for ground water pollution.

With the declared objective of providing at least the basic amenities there has been a tremendous development in India, in the agricultural sector, with concomitant pressure on the water resources. The waste generated by anthropogenic activities has not only polluted the environment as a whole but had a particular detrimental effect on the quality of equation-environment too. Leachates from composts pits animal. Refuse of garbage of dumping grounds nutrients enriched return irrigation flows, seepage from septic tanks, seepage of sewage, etc. has adversely affected the ground water quality in several parts of India.

Industrial waste water is formed at industrial plants, washing and rinsing of equipment, rooms, etc. these operations results in the pollution of the near of aquatic system because some of the products and by products are discharged, either deliberately or intentionally in to them.

Kaushik [7] analyzed water of 100 wells from 64 villages located in Delhi and found that the total alkalinity varied between 130 to 740 mg/L.

The concentration of chloride, sulfate and nitrate ranged between 12 to 1280 mg/L, 3.5 to 760 mg/L and 1 to 376 mg/L, respectively.

Olaniya [12] carried out a survey of wells and tube wells for the supply water in the city during year 1967–68. The physico-chemical characteristics of these wells indicated that well water was suitable away from city. The analysis of sample revealed that some of the physico-chemical properties of water are beyond the permissible limits prescribed for domestic consumption.

Schmidt [16] studied the nitrate nitrogen concentration in the vicinity of fresno clovis areas in USA. The scientist has reported that the level of nitrate exceeded 455 mg/L at least once in 1947. The distribution of well was very wide as twelve of the wells were in municipal area and remaining were in the agricultural land.

Handa [4] studied the nitrate content of ground waters in several parts of India. The analysis of data revealed the nitrate content of surface water was quite high. Nearly 50 of stream water analyzed had less than 3.0 mg/L N03. Only in about 10 stream water analyzed, N03 content exceeded 10 mg/L of nitrate for dug well or water table aquifer, however a different pattern was observed; a considerable percentage of well water analyzed had over 100 mg N03 per liter. The examination of data for Uttar Pradesh showed that only 40 of dug well water analyzed had less than 40 mg N03 per liter, however, a few were found to contain even more.

Shrivastava et al. [21] studied changes in water quality along the River, polluted by effluents from the Orient Paper Mill, Amali in Madhya Pradesh. The water sample of 15 stations, covering a distance of about 216 km along the river course, were subjected to hierarchical cluster analysis to determine similarity relations with respect to seven water quality parameters

Mahayana and Suresh [10] analyzed 56 samples of groundwater for pH, specific conductance, hardness, Na, K, Ca, Mg, Cl, HCO_3, and SO_4. Most of these chemical parameters were found correlated with one another. The result indicated that majority of ground water samples were acidic in nature, i.e., ranged from 3.1 to 7.5. The acidic nature may attribute to the CO_2 that is incorporated in to ground water by bacterial oxidation. Concentration of Na and Ca varied in the range of 1–117 mg/L and 2–74 mg/L, respectively. The study further revealed that four type of groundwater having different chemical parameters combination Na-Ca-Cl, Na-Ca-CO$_3$, Na-Ca-Cl-HCO$_3$ and Ca-Mg-Cl in Mangalore and the city had two distinct ground water zones namely Na-Ca-Cl and Na-Ca-HCO$_3$.

Sahgal et al. [14] studied the nitrate pollution of ground water in Lucknow area. The study was undertaken keeping the great importance of nitrate in water supplies arising with the establishment of a direct relationship between nitrate Concentration in drinking water and the incidence of methemoglobinemia, which can fatal to infants. The analysis of water samples demonstrated that nitrate levels several times higher than prescribed limits of 45 mg/L were present. Nitrate concentration in ground water exceeded 100 mg/L at several places and concentration as high as 650 mg/L had been observed in shallow ground water. Pollutants had adversely affected the ground quality in the depth range of 100–125 m with nitrate concentration of 145,168, and 590 mg/L in some of the tube wells.

Faillat and Rarraud [3] studied a number of wells dug in most French speaking countries in West Africa which were being used to supply rural population with potable understand water through village hydraulic program me. Most of these wells were located in salty sandstone like rocks that lie beneath 5–40 m of weathered and decayed material. Most of these wells had total depth of 40–0 m, a daily discharge of about 10m^3, and instantaneous discharges 1–5 m^3/ha. In spite of favorable hydro geological and environmental condition, many of these wells displayed high nitrate contents. A statistical approach on a regional scale was used to determine the cause of the high nitrate contents. It was determined that the source of nitrates is linked mainly with deforestation by man.

Salade and Sims [15] reported that N$_2$ from poultry manure (PM) and ammonical fertilizers undergoes rapid nitrification upon addition to soils, making it highly susceptible to leaching and thus creating a potential for ground water contamination by NO$_3$-N. Any management technique that could delay nitrification could be desirable.

Tase [25] mentioned that normal physical condition in Japan restricted the possibilities of ground water contamination; human activities are threatening the ground water resources. A survey conducted by the environmental agency of Japan showed nationwide spreading of organic substances, such as tri-chloro-ethylene as well as nitrogen compounds. Synthetic had also been detected even in rural areas and in deep confined aquifer, although there concentration was not high.

Somasundaram et al. [23] carried out a study to gain an idea of the inorganic quality of water in the wells penetrating the shallow (20 m thick) alluvial aquifer below the city of Madras. Two general surveys of selected wells across the city indicated that relative to India domestic water standard, of the 93 ground water samples obtained, 25 exceeded the Ca limit, 11% the Mg limit, 43 the total dissolved solids limit, 40 the SO_4 limit and 70 the NO_3 limit. As an indicator of biological contamination, the high concentration of NO3 (0 to 1040 mg/L) were particularly worrying. A detailed survey of sites close to a city river was also undertaken. High heavy metal concentration were detected, with Arsenic up to 0.42 mg/L, mercury to 0.02 mg/L, lead to 1.82 mg/L and Cadmium to 1.31 mg/L. Microbes have been found in several of the wells.

Kumar [8] collected water samples from different areas of *Tarai* and *Bhabar* region of Nainital district and analyzed those samples for pH, color, conductance hardness, Bio-chemical Oxygen Demand (BOD), Chemical Oxygen Demand (COD), Dissolved Oxygen (DO), alkalinity and chloride with special reference to nitrate. It was observed that pH was ranging 5.7 to 8.1, EC 195 to 780 μmhos, DO 0.8 to 8.9 mg/L, BOD 0.8 to 1733 mg/L, COD 11.35 to 4750 mg/L and alkalinity varied from 26 to 498 mg/L. The review shows various relevant aspects of ground water pollution.

Chadha and Tamta [2] studied the causes and nature of various inorganic, organic and metallic constituents in the wells in major cities in India. The inorganic, organic and heavy metal constituents including colors unit frequently exceeded WHO/BIS standards for drinking water. Ground water chemical constituents were measured over a field season program in ministering well placed strategically in relation to city waste drainage, industrial drains and spread. Result indicated that rapid urbanization, industrialization was major source of the organic and heavy metals in ground waters and inputs are markedly dependant on wet drains, resulting in surge of the chemical constituents in the ground water.

Pandey [13] studied the pollution extent in surface water, bed sediments and ground water in bed width of Ramganga river for a stretch of 36 Km in Moradabad district of Uttar Pradesh. The untreated water effluents wastes of nearly 450 electroplating plants and entire brass and stainless steel industry apart from the domestic wastewater mainly cause the pollution. The physical, chemical and biological parameters were determined. The study indicated increase in pollution level along the downstream stretch of river under consideration.

Sharma and Chauhan [19] studied the detrimental effects of environmental pollution in the *Tarai* belt of Udham Singh Nagar district, during the last four decades. Due to cleaning of dense forests, urbanization, industrialization and agricultural activities, there have been undesirable changes in ground water quality. The physico- chemical properties of the ground water showed that it was most suitable for irrigation and industrial uses and unsuitable for drinking purpose due to high concentration of coliform bacteria.

Bala and Rawat [1] studied the nitrogen concentration in ground water in Udham Singh Nagar district. Analysis of water of Gola river, ground water along the bank of Gola river was done. 19 samples were collected form different locations. Study revealed that water taken from ground water of shallow aquifer is not safe for consumption of human beings, without proper treatment of coliform bacteria. Chloride content of all samples was within permissible limit for drinking purpose.

Striger et al. [24] developed a simple multivariate analysis based methodology to create Ground Water Quality Index (GWQI) and a composition index (GWCI), with the aim of monitoring the influence of agriculture on several key parameters of ground chemistry and pot ability. The methodology was based on the definition of two standard water samples of high and low quality hate, together with the actual data, were run through a statistical algorithm known as correspondence factor analysis and communication tool evaluated case studies in the south of Portugal. Index maps were created to provide a comprehensive picture of the contamination problem and easy interpretation for people outside the scientific domain. In the case studies, the GWQI maps revealed that groundwater quality in the upper aquifers was extremely low, with an almost complete absence of potable water. The impact of agricultural activity on the ground water composition showed a large partial variability was related to crop type and aquifer lithology. The above studies showed that lot of work has been done to industrial effluent and their impact on the plant growth. Such studies were area specific or for particular industry in that area. No work could be available on the study of pollution of ground water resources due to fertilizer factory in North West Uttar Pradesh and the effect of such fertilizer factory effluent on plant growth and crop yield. Keeping in view the present study was taken up to asses impact of effluent of IFFICO, Anola on water resources of the nearby and on growth and yield of wheat crop.

Hariharan [5] studied the water quality of Vuda (Mithilapuri) Colony of Vishakhapatnam (Andhra Pradesh). Physico-chemical analysis of well and bore well water samples was carried out from different station of area during the month of September 2004. The result indicates that the water is extensively hard and the reason might be sewage pollution through the sandy aquifer, which is likely to be influenced by salt water contamination.

Singh et al. [22] studied the change in the shallow water quality during one decade, 1994–2007, in the *Tarai* of Udham Singh Nagar and Nainital districts of Uttarakhand. The study revealed that the ground water is not safe for human consumption due to presence of coliform bacteria which indicated the presence of pathogenic bacteria. The total dissolved salts were found within the safe limits for drinking purpose for all the samples except the water collected from Central Institute for Medicinal and Aromatic Plants (CIMAP) and railway station, Pantnagar. The chemical oxygen demand of water samples has increased over a decade due to increase in concentration of inorganic chemicals due to industrial effluent of Century Paper Mill, Lalkuan (Nainital).

Lakhera and Singh [9] studied the effect of monsoon on water quality of shallow aquifer in *Tarai* of Udham Singh Nagar district. The study revealed that the ground water is not safe for human consumption due to presence of coliform bacteria, which indicated the presence of pathogenic bacteria. Water from the entire study area is suitable for irrigation purpose as the parameters important for irrigation suitability are within prescribed limits.

Sharma and Sharma [20] studied ground water quality of selected villages of Alwar district, Rajasthan. The result revealed that Electrical Conductance (EC) (500 to 8300 μmhos/cm) and alkalinity (200–1048 mg/L) of all samples were very high as compared to permissible limit, which can be correlated with high fluoride total hardness and dissolved solids. More than 1.5 mg/L concentration of fluoride was observed in 18.2% sample.

Where 31.8% water samples contain high concentration of nitrate and 86.3% samples contain high total hardness compare to permissible limits. The results suggest that quality of ground water of villages of Alwar district is unfit for human consumption, as level of fluoride, EC and alkalinity is very high which may lead to various health problems.

Hiremath et al. [6] analyzed physico-chemical parameters of ground water of municipal area of Bijapur (Karnataka) to study the quality of water and suitability for domestic purpose. The parameters: pH, EC, TDS, Turbidity, Total hardness and content of Fluoride, Sulphate, Chloride were studied and compared with the standard values prescribed by ICMR, WHO and APHA. The investigation revealed that the quality of water of a source varies from season to season and some of the water samples are unfit for drinking and utility purpose.

Nagarnaik and Bhalme [11] analyzed the water samples taken from different areas (Sindi-Meghe, Ram Nagar, Gopuri, Boregaon-Meghe, Sawangi-Meghe, Mhada Colony, Karala square, Shastri square) of Wardha City. Based on different parameter pH, temperature, total dissolved solids, alkalinity, hardness, suspended solid, dissolved solid, chloride, turbidity, ions, and MPN, the water found unfit for drinking purpose.

Sethi [18] analyzed 20 water samples from dam for pH, electrical conductivity, total hardness, Ca, Mg, total alkalinity, chloride, salinity and Fluoride. Twenty samples of water from dam ponds were analyzed. pH was measured by pH Meter; EC was measured by Conductivity Bridge. Total hardness, calcium, magnesium, total alkalinity, chloride, salinity were determined by titration method. The dam water was found safe for drinking as well as for irrigation purpose.

Sehgal [17] in the study of the quality of drinking water in Dhakuakhana Sub division of Lakhimpur district, Assam, India analyzed 30 water samples for pH, total hardness, fluoride, nitrate, arsenic, sodium, potassium and iron by using standard methods and found that concentration of all parameters were within permissible limit.

Review of studies carried out by different investigators of various agencies indicates that there is wide spread groundwater pollution in the country. Ground water in several areas where sewage is being discharged without proper treatment has been adversely affected by contaminants associated with sewage. High level of potassium and phosphate have been reported in ground water from several places in Punjab, Haryana and Uttar Pradesh. Groundwater is moderately to highly saline in several parts of Rajasthan, Punjab, Gujarat, Haryana, Delhi and many other areas.

8.3 MATERIAL AND METHODS

8.3.1 DESCRIPTION OF AREA

The study area, i.e., the area adjoining Sitarganj, Shakti Farm, Shahdora and Pulbhatta, is spread in the inter-basin of Saryu and Gola river, it is also located between three major water reservoirs named Baighul reservoir, Dhora reservoir and Nanak Sagar reservoir in Udham Singh Nagar district of Uttarakhand.

8.3.2 CLIMATE

The climate of study area is marked by extreme winters and very hot summers. The temperature during winters nearly drops to the freezing point. The maximum temperature in summers occasionally reaches to 46°C. Normally minimum rainfall is observed in the months between November to March, maximum rainfall is observed during monsoon period, July to mid-September the average rainfall observed in the area is of about 1122 mm. The mean winds speed during winter season varies from 2.6 km/hr to 5.3 km/hr during summer; it varies from 5 km/hr to 7.6 km/hr in general. Relative humidity is' high in winters (December and January) because of low atmospheric temperature and in rainy month (July to September) when high temperatures are more compensated by heavy rainfall. The periods of low humidity coincide with dry months.

8.3.3 TOPOGRAPHY

The topography of study area has elevation of 298 meters (978 feets) from sea level, study area has variation from *Bhabar* to *Tarai*. *Tarai* occurs south of *Bhabar* with gentle slope towards south. The maximum width of

Tarai belt is encountered in Kashipur, Nagla, and Khatima section and is about 26 km in north and south.

The *Tarai* formation, also of recent formation consist of clays, sandy clays sands and occasionally gravel built up and deposited in number of strata overlying the sub surface continuation at Shivalik hills formation, the top layer of 50 cm consist of fine sand and -clay, beneath which lie alternating layers of coarse material and fine material of varying thickness. The layers with coarse texture, is potential confined water bearing strata, probably all are hydrologically interconnected and receive water from precipitation on Shivalik mountain range. The northern boundary of *Tarai* formation is not clearly defined, as it merges gradually with the gangetic alluvium, and gradually taken the zone where the flowing conditions cease to exist in tube wells.

8.3.4 COLLECTION OF SAMPLE

The water samples for physico-chemical analysis were collected from hand pump of twenty locations (Table 8.1) of study area. Samples were collected using plastic bottles and were kept in incubator so that no or minimum changes occur in physico-chemical characteristics of the water samples.

TABLE 8.1   Locations of Sampled Hand Pump in the Study Area

Sl. No.	Location	Sl. No.	Location
1	Shadhaura (Mujib Sweet Shop)	11	Govind Nagar (Pandav Gaon)
2	Shadhaura (M.K Book Depot)	12	Kalyanpur
3	Shamalpura (Bengali Colony)	13	Sisona (Kumaon Furnitures)
4	Shakti Farm (Beema Seva Kendra)	14	Sisona (Bartuabag)
5	Shakti Farm (GIC, Campus)	15	Sisaiya (Taneja Gen. Store)
6	Shaktigara (Shastri ward 2)	16	Sirga (Shakti Farm T point)
7	Shakti Farm (Main Market)	17	Uttam Nagar (Gurudwara Buddha Sahib)
8	Shakti Farm (Tagore Nagar)	18	Shankar Farm
9	Vaikanthpur (Number 1)	19	Shankar Farm (Gutiya)
10	Vaikhantpur (Panch Quarter)	20	Pulbhatta

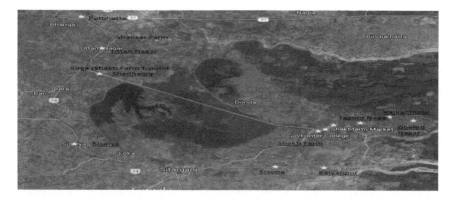

FIGURE 8.1   Location of sampled hand pumps in the study area.

Map of Study Area shows the study area and various location from where water samples has been collected for the physico-chemical analysis.

8.3.5   PHYSICO-CHEMICAL ANALYSIS

8.3.5.1   Odor, Color, Temperature and pH

The odor was measured by inhaling; color was identified by visual interpretation. The cap of electrode of pH meter was removed and the electrode was dipped into water sample and few minutes time was allowed until reading is stabilized. The reading shown on the display of pH meter was pH of sample.

8.3.5.2 Turbidity

The turbidity was determined with the help of Multi-parameter water Quality Instrument known as (TROLL 9500). The turbidity measurement is a default. Measurement of the equipment with the help of the electrode.

8.3.5.3 Acidity

The measurement of acidity in the water samples was carried out by the standard analytical method. The reagents used in the analytical method were Phenolphthalein indicator solution, which has prepared by dissolving 500 mg Phenolphthalein in 500 mL ethyl or isopropyl-alcohol and 50 mL distilled water; Methyl Orange indicator (0.5) obtained by dissolving 0.5 mg methyl orange in 100 mL distilled water; Sodium hydroxide (0.05 N) prepared by adding 40 g NaOH in distilled water and making the volume to 1000 mL. Further 50 mL of this solution were diluted to 1000 mL, to get a solution of 0.05 N for measurement of acidity of the water 10 mL of color less sample was taken in a conical flask. Three drops of Methyl indicator were added to it. The sample was titrated with 0.05 N solution of Sodium Hydroxide until the color change to faint pink at the end. At the end point three drops of phenolphthalein indicator was added to it and continued until the content turned pink. The calculations are performed by using the following equation.

$$\text{Total Acidity} = \frac{\text{Volume of NaOH} \times 50 \times 1000}{\text{Volume of the sample taken for titration}} \qquad (1)$$

8.3.5.4 Free CO_2

A volume of 100 mL of water sample was taken in a conical flask and 3–4 drops of phenolphthalein indicator was added to it. The sample was titrated with 0.05 N NaOH until pink color appeared at the endpoint. Free carbon dioxide was calculated using following equation.

$$\text{Free Carbon Dioxide} = \frac{\text{Volume} \times \text{Normality of NaOH} \times 44 \times 1000}{\text{Volume of the sample taken for titration}} \qquad (2)$$

8.3.5.5   Conductance

The electrical conductivity was measured by pocket EC tester (conductance meter). The EC tester was set ON and cap of electrode of was removed to dip the electrode into the water sample and few minutes time was allowed to the tester to stabilize. The reading shown on the display of conductance meter was taken as conductance of the sample.

8.3.5.6   Nitrate

The nitrate-nitrogen was measured by using Multi parameter Water Quality Instrument (TROLL9500). The nitrate sensor was inserted on the marked port of the instrument. The instrument was first calibrated for low, medium and high concentration of NO_3 by known values of the ammonium nitrate solutions. After calibration the sensor was dipped into the water sample of unknown nitrate concentration and the reading from the display of the instrument, was noted down as nitrate content of water sample.

8.3.5.7   Calcium, Magnesium Content and Hardness

Hardness is an indication of the amount of salts of calcium and magnesium in the water. Calcium and magnesium are essential elements for the plant growth that are reported in parts of element per million parts water (ppm) on the weight basis. Calcium in the range of 40–100 ppm, and magnesium in the range of 30–50 ppm are considered desirable for irrigation water. The classification of water on the basis of hardness, as $CaCO_3$ is given in Table 8.2.

TABLE 8.2   Classification of Water on the Basis of Hardness (as CaCO3, mg/L)

Classification	Hardness (as $CaCO_3$, mg/L)
Soft	0–17.1
Slightly hard	17.1–60
Moderately hard	60–120
Hard	120–180
Very hard	180 and above

The Ca, Mg content and hardness was measured through standard analytical methods. Ethylene-Di-amine Tetra Acetic acid (EDTA solution) (O.01N) was prepared by dissolving 3.723 g of disodium salt of EDTA in distilled water to prepare one liter of solution. The prepared solution was stored in polyethylene bottle. Sodium Hydroxide (1N) solution was obtained by dissolving 40g of NaOH in distilled water to make one liter of solution. Murexide oxide indicator was prepared by mixing 0.2 g of ammonium purpurate in 100 g of NaCl and grinding it well. Sodium sulfide solution was prepared by dissolving 5 g of hydrated. sodium sulfide ($Na_2S.9H_2O$) in 100 mL distilled water. Sodium sulfide solution was stored in tightly closed bottle to prevent oxidation. Buffer solution was obtained by dissolving the two solutions A and B. The solution A was made by dissolving 1.9 g NH_4Cl in 143 mL of concentrated NH_4OH. The solution B was obtained by dissolving 1.179 g di-sodium EDTA and 0.78 g of Magnesium Sulphate hepta hydrated in 50 mL of distilled water.

Both the solution were mixed and diluted to 250 mL with distilled water. Erichrome Black-indicator was prepared by mixing 0.01 mg of Erichrome Black-T in 100 mg NaCl and grinding it.

8.3.5.7.1 Calcium content

First 10 mL of sample was taken and then diluted to 50 mL by adding distilled were 0.2 mL of NaOH sol. (1N) and 0.2g Murexide indicator were added to it. Solution was titrated against EDT A sol. until pink color changed to purple. The content of Ca in sample is calculated by:

$$Ca\,(\,mg/l) = \frac{Vol.\ of\ EDTA\ used \times 400.8}{Volume\ of\ sample\ taken\ for\ titration} \tag{3}$$

8.3.5.7.2 Total hardness

First 10 mL of sample was taken in a conical flask and diluted to 50 mL. Then 1 mL of buffer sol. and 2 g of Erichrome Black-T indicator were added to it, the color of solution was changed to wine red. The contents were then titrated against EDTA solution until the wine red color changes to blue at the end point.

The value of total hardness, calcium hardness and magnesium hardness were calculated using following equations:

$$\text{Total hardness(as CaCO}_3) = \frac{\text{Vol. of EDTA used} \times 1000}{\text{Volume of sample taken for titration}} \quad (4)$$

8.3.5.7.3   Calcium hardness

Calcium hardness as $CaCO_3$ (mg/L) = Calcium content (mg/L) 2.497

8.3.5.7.4   Magnesium hardness

Magnesium hardness as $CaCO_3$ (mg/L) = (Total hardness – Calcium hardness)

8.3.5.7.5   Magnesium content

Magnesium, mg/L = magnesium hardness x 0.244

8.3.5.8   Chloride Content

Although chloride is essential to the plant in very low amounts, it can cause toxicity to sensitive crops at high conc. like sodium, high chloride concentration case more problems when applied with sprinkler irrigation. Leaf burn under sprinkler from both sodium and chloride can be reduced by night time irrigation or application on cool, cloudy days.

The effect of chloride concentration on the crops can be classified as follows:

Chloride (mg/L)	Effect on crops
Below 70	Generally safe for plant
70–140	Sensitive plants shows injury
141–350	Moderately tolerant plants show injury
Above 350	Can cause severe problems

The chloride content was measured through standard analytical methods by using 0.02 N Silver Nitrate and 5 Potassium Chromate solutions.

Silver Nitrate (0.02N) was prepared by dissolving *3Ag* of AgNO₃ in distilled water and diluting it to 1000 mL. Because of the oxidizing properties of the silver nitrate the solution was kept in dark (Amber) bottle. Potassium Chromate solution (5) was obtained by dissolving 5 g of Potassium Chromate in 100 mL of distilled water.

About 50 mL of water from sample was taken into a flask and 2 mL of Potassium chromate solution was added to it. Then this solution was titrated against 0.02N Silver Nitrate solution until persistent red tinged appeared. The value was calculated using following equation.

$$\text{Chloride(mg/l)} = \frac{(\text{Volume and Normality) of AgNO3} \times 35.5 \times 1000}{\text{Volume of sample taken for titration}} \tag{5}$$

8.3.5.9   Measurement of Alkalinity

The alkalinity was estimated through standard method. Hydrochloric acid solution (O.O1N) was prepared by diluting 12N concentrated HCl (specific gravity 1.18) to 12 times to prepare in HCl and diluted again to 10 times to make 0.1N HCl and standardized it against sodium carbonate solution. Methyl orange indicator soln., 0.05 was prepared by dissolving 0.5 g of methyl orange in 100 mL of distilled water.

Phenolphthalein indicator solution can be prepared by dissolving 0.5 g of phenolphthalein in 50 mL of 95 ethanol and 50 mL of distilled water was added to it. Add 0.05N carbon dioxide free NaOH drop wise until the solution turned faint pink. Sodium carbonate solution; O.IN was obtained by dissolving 5.3 g of sodium in distilled water to prepare 100 mL of sol.

The 100 mL of sol. was taken and 2 drops of phenolphthalein indicator was added to it. If color changed to pink, titration was done 0.1N Hel until the color disappeared at the end point, then 3 drops of methyl orange indicator were added to and titration was continued until the yellow color changed to pink at end point. For samples which did not changed color after addition of phenolphthalein, 3 drops of methyl orange indicator were added and titrated with O.IN Hel until the yellow color changed to pink at the end point. The value was calculated using Eq. (6):

$$\text{Total alkalinity(as CaCO}_3\text{)(mg/l)} = \frac{(A \times \text{Normality) of HCl} \times 50 \times 1000}{\text{Volume of sample taken for titration}} \tag{6}$$

where A = mL of total HCl used with phenolphthalein and methyl orange indicators.

8.3.5.10 Measurement of Ammonia

The ammonia was measured by using Multi-Parameter water quality Instrument (TROLL 9500). The NH_4 sensor was inserted at the marked port of instrument. After calibration for low, medium and high concentration of NH_4, the sensor was dipped into water sample of unknown ammonium concentration and the reading from the display of the instrument, was noted down as concentration of Ammonia in the sample.

8.3.5.11 Sodium

The amount of sodium is less than 200 mg/L. sodium is not considered a toxic metal, and normal adults may consume 5,000 to 10,000 milligram per day without any adverse effects. The average intake of sodium from water is only a small fraction of that consumed in normal diet. The recommended maximum level of people suffering from certain medical conditions such as hypertension, congestive heart failure or heart disease is 20 mg/L. If in doubt, consult a physician. Sodium is a significant factor in assessing water for irrigation and plant watering. High levels affect soil structures and a plant's ability to take water.

The sodium content in water samples is estimated by using Flame Photometer in water quality testing lab of the department.

8.3.5.12 Potassium

There is no guideline or recommended limit for potassium in water. Water softeners that regenerate using potassium chloride can significantly raise the level of potassium in water. It is recommended that people with kidney diseases or other conditions such as heart diseases, coronary artery disease, hypertension, diabetes and those who take medication that interferes with how the body handles potassium do not drink water from a water softener that use potassium chloride.

The potassium content in water samples is estimated by using Flame photometer in the water quality testing lab of the department.

8.3.6 IRRIGATION WATER QUALITY

The parameters which determine the irrigation water quality are divided into three categories: physical, chemical and biological. The chemical characteristics of irrigation water refer to the content of salt in the water as parameters derived from the composition of salts in the water, parameters such as Electrical Conductivity/Total Dissolved Solids (EC/TDS), Sodium Adsorption Ratio (SAR) alkalinity and hardness. The main problem related to irrigation water quality is the water salinity. Water salinity, refers to the total amount of salts dissolved in the water but it does not indicate which salts are present in it.

High level of salts in the irrigation water reduces water availability to the crop (because of osmotic pressure) and cause yield reduction. Above a certain threshold, reduction in crop yield is proportional to the increase in salinity level. Different crops vary in their tolerance to salinity and therefore have different thresholds and yield reduction rates.

The most common parameters used for determining the irrigation water quality, in relation with its salinity, are EC and TDS. The suitability of an irrigation water depends upon several factors, such as, water quality, soil type, plant characteristics, irrigation method, drainage, climate and the local conditions. The integral effect of these factors on the suitability of irrigation water (SI) can be expressed by relationship given below:

Suitability of irrigation water, SI = $\int QSPCD$.

where, Q = quality of irrigation water, that is total salt concentration, relative proportion of cations, etc.; S = soil type, texture, structure, permeability, fertility, calcium carbonate content, type of clay minerals and initial level of salinity and alkalinity before irrigation; P = salt tolerance characteristics of the crop to be grown, its variety and growth stage; C = climate, that is total rainfall, its distribution and evaporation characteristics; and D = drainage conditions, depth of water table, nature of soil profile, presence of hard pan or lime concentration and management practices.

Irrigation, water can be classified in five classes depending upon its chemical properties, as given in Table 8.3.

TABLE 8.3 Guideline for Evaluation of Irrigation Water Quality

Water class	Sodium (Na), %	Electrical conductivity (µS/cm)	SAR	RSC, meq/L
Excellent	<20	<250	<10	<1.25
Good	20–40	250–750	10–18	1.25–2.0
Medium	40–60	750–2250	18–26	2.0–2.5
Bad	60–80	2250–4000	>26	2.5–3.0
Very bad	>80	>4000	>26	>3

Source: (BIS 11624:1986).

8.4 RESULTS AND DISCUSSION

The study was conducted for the regions of Shahdaura, Shemalpura, Shakti Farm, Shaktigara, Vaikutpur, Govindnagar, Kalyanpur, Sisona, Sisaiya, Sirga, Uttam Nagar, Shankar Farm and Pulbhatta. The samples collected from identified locations were tested for 18 parameters using standard methods, in Water Quality and Pollution Control Laboratory, Department of Irrigation and Drainage Engineering. The values of different parameters obtained from lab analysis of the samples are shown in Table 8.4. The desirable and permissible limits of water quality parameters in details for various uses as suggested by BIS has been reproduced in appendices the result analysis of the present study are discussed parameter wise as given in the following subsections.

8.4.1 COLOR, ODOR, AND PH

The samples are mostly odorless and colorless except samples collected from the hand pumps located at Viakanthpur no.1 and Uttam Nagar near Gurudwara Buddha Sahib. The color of water from these locations was having the light yellow color. The light yellow color may be due to the higher concentrations of total dissolved solids and iron, in water samples.

The pH of the water sample was found in the ranges 6.3 at Uttam Nagar to 7.5 at Shakti Farm, Tagore Nagar. According to WHO (1992) Standards, best and ideal pH value for human consumption is 7.0, but it may vary from 6.8 to 8.5. Hence the water from other places except Vaikanthpur

Table 8.4a Physico-Chemical Characteristics of the Water Samples

Properties	Location of sampled hand pump									
	Shadaura (Muljee Sweet Shop)	Shadaura (M.K book depot)	Shamalpura (Bengali Colony)	Shakti Farm (Beema Sewa Kendra)	Shakti Farm (CGIC)	Shaktigara (Shastri Ward 2)	Shakti Farm (Main Market)	Shakti Farm (Tagore Nagar)	Valkanthpur (Number 1)	Valkhant-pur (Panch Quarter)
Odor	Odorless	Odorless	Odorless	Odorless	Odorless	Odorless	Odorless	Odorless	Objec-tionable	Odorless
Color	Colorless	Colorless	Colorless	Colorless	Colorless	Colorless	Colorless	Colorless	Light yellow	Colorless
Acidity (mg/L)	55	20	32.5	20	27.5	25	25	30	75	17.5
Alkalinity (mg/L)	485	285	340	225	210	230	235	345	500	190
Ammonium (mg/L)	0.01	0.06	2.65	5.68	6.3	5.96	0.6	0.04	0.02	0.01
Ca (mg/L)	19.23	16.03	13.62	14.42	12.82	14.42	16.83	17.64	18.43	8.02
Ca hardness (mg/L)	48.02	40.03	34.01	36.01	32.01	36.01	42.02	44.03	46.02	20.02

Parameter										
Chloride (mg/L)	32.66	8.52	9.96	4.26	7.1	14.2	4.26	18.46	66.74	5.68
EC (µS/cm)	650	460	600	400	410	400	400	470	850	330
Free CO_2 (mg/L)	35.2	19.8	26.4	19.8	24.2	22	15.4	46.2	72.6	13.2
Iron (mg/L)	1.15	0.6	0.35	0.75	0.065	0.8	0.04	5.5	9.15	0.05
Mg (mg/L)	15.31	8.77	13.26	14.05	11.99	11.60	7.60	9.84	13.07	10.73
Mg hardness (mg/L)	62.77	35.97	13.26	57.58	49.16	47.58	31.17	40.37	53.57	43.98
Nitrate (mg/L)	0.77	1.1	1.2	0.77	0.77	0.87	0.5	0.8	1.05	1.26
pH	7.3	6.9	6.9	6.8	6.6	6.8	6.8	7.5	6.5	6.8
Potassium (mg/L)	24	21	22	19	20	18	19	24	21	14
Sodium (mg/L)	10	14	12	15	16	14	13	14	19	8
TDS (mg/L)	440	310	400	270	280	270	270	310	570	220
Total hardness (mg/L)	82	52	68	72	62	62	48	58	72	52
Turbidity (NTU)	10	3.6	1.8	0.7	0.5	1.2	0.4	0.2	5	0.1

Table 8.4b Physico-Chemical Characteristics of the Water Samples (contd.)

Properties	Location of Sampled Hand Pumps									
	Govind Nagar (Panday Gaon)	Kalyanpur	Sisona (Kumaon Fur-nitures)	Sisona (Bartua-bag)	Sisaiya (Taneja Gen. Store)	Sirga (Shakti Farm T-Point)	Uttam Nagar (Gurudwara Buddha Sahib)	Shankar Farm	Shankar Farm Gutiya	Pulbhatta
Odor	Odorless	Odorless	Odorless	Odorless	Odorless	Odorless	Objectionable	Odorless	Odorless	Odorless
Color	Colorless	Colorless	Colorless	Colorless	Colorless	Colorless	Light yellow	Colorless	Colorless	Colorless
Acidity (mg/L)	55	47.5	65	70	67.5	57.5	145	55	72.5	30
Alkalinity (mg/L)	390	395	320	345	480	420	730	485	585	385
Ammonium (mg/L)	0.01	0.6	0.78	58.14	3.3	0.04	27.1	10.65	0.4	0.07
Ca content (mg/L)	20.84	24.84	20.04	24.84	16.03	16.83	20.04	21.64	26.45	25.65
Ca hardness (mg/L)	52.04	62.03	50.04	62.03	40.03	42.02	50.04	54.04	66.05	64.05
Chloride (mg/L)	4.26	26.98	14.1	42.6	71	18.46	217.26	45.44	78.1	7.1
EC (μS/cm)	620	650	530	830	1320	620	1520	730	1030	660

Parameter										
Free CO$_2$ (mg/L)	55	55	44	55	70.4	41.8	127.3	55	88	30.8
Iron (mg/L)	0.7	1.1	2.6	1.55	0.95	0.7	1.6	1.3	0.7	0.25
Mg (mg/L)	8.58	19.32	12.68	13.46	20.48	17.36	32.69	17.65	27.78	12.29
Mg hardness (mg/L)	35.16	79.16	51.96	55.16	83.97	71.17	133.9	72.36	101.5	50.35
Nitrate (mg/L)	1.3	0.8	0.9	0.85	0.5	1	1.87	0.82	1.2	0.99
pH	6.7	6.5	6.5	7.1	6.6	6.9	6.3	7.2	6.4	7.2
Potassium (mg/L)	17	22	21	20	21	22	23	14	21	18
Sodium (mg/L)	11	18	16	22	18	14	21	8	16	12
TDS (mg/L)	420	440	360	560	890	420	1020	490	690	440
Total hardness (mg/L)	56	104	72	80	100	88	154	94	128	76
Turbidity (NTU)	3.5	0.7	1	0.8	4.4	1.2	0.2	1.2	0.4	0.6

no.1 (6.5), Uttam Nagar (6.3), Sisaiya (6.6), Sisona (6.5), Shankar Farm Gautiya (6.4), Shakti Farm (GGIC) (6.5) and Govind Nagar (6.7) were not safe for human consumption from the point of view of pH.

The normal range for the irrigation use is from 6.5 to 8.4, so samples from Uttam Nagar (6.3) may not be suitable for irrigation purpose.

8.4.2   TOTAL DISSOLVED SOLIDS (TDS)

The TDS of water samples was found in the range of 220 to 1020 mg/L. The highest TDS value was 1020 mg/L was observed in the samples collected from Uttam Nagar and lowest in Vaikanthpur (Panch Quarter) 220 mg/L. For the drinking purpose the TDS must be in between 500 to 2000 mg/L which shows water sample collected from Shakti Farm (Tagore Nagar), ShankarFarm, Sahdaura, Sisona, Shakti Farm (GGIC), Kalyanpur, Shakti Farm (Bema Sewa Kendra) Sisona, Shaktigarh, Sirga, Pulbhatta, Vaikanthpur (Panch Quarter), Shakti Farm (Main Market), Shamalpura are suitable for the drinking purposes.

8.4.3   CHLORIDE

The concentration of chloride varied between 4.26 mg/L to 217.26 mg/L. The minimum value of 4.26 mg/L was observed in Shakti Farm (Bema Sewa Kendra) and maximum concentration of 217.26 mg/L was observed in Uttam Nagar. As per BIS standards the permissible value of chloride for domestic purpose is 200 to 600 mg/L. The results indicate that water from all locations is safe for drinking. The concentration of chloride is found less than 200 mg/L at all locations except Uttam Nagar.

8.4.4   ALKALINITY

The alkalinity of the samples varied from 190 to 730 mg/L. as per BIS the value of alkalinity be in the range of 200 mg/L to 550 mg/L. The alkalinity of water from Uttam Nagar (730 mg/L), Shankar Farm Gutiya (585 mg/L) may lead to the corrosion and is not safe for the drinking purposes due to higher concentration of bi-carbonates. This is due to dissolution of Carbon dioxide and alkalinity imparts bitter taste.

8.4.5 ACIDITY

The acidity of sample varied between 17.5 mg/L at Vaikanthpur (Panch Quarter) to 145 mg/L at Uttam Nagar. The reason of higher acidity is the higher concentration of chloride in water.

8.4.6 FREE CARBON DIOXIDE

The Carbon dioxide level varied from 13.2 mg/L at Vaikanthpur (Panch Quarter) to 127.6 mg/L at Uttam Nagar. Uttam Nagar reading is showing that the maximum dissolution of carbon dioxide in it. The higher value of free carbon dioxide in water makes it bitter in taste.

8.4.7 TOTAL HARDNESS

In ground water hardness is mainly due to carbonates, bicarbonates, chlorides, sulfates of calcium and magnesium. Data revealed that value of hardness are ranged between 48 mg/L Shakti Farm (Main Market) to 154 mg/L in Uttam Nagar. The permissible limit of total hardness is 100–500 mg/L. The water is safe for domestic use in all the locations.

8.4.8 CALCIUM HARDNESS

The calcium hardness of the sample varied from 20.01 mg/L at Vaikanthpur (Panch Quarter) to 66.04 mg/L at Shankar Farm Gutiya, which is well in permissible range of 75 mg/L to 200 mg/L.

8.4.9 MAGNESIUM HARDNESS

The Mg hardness varied from 31.17 mg/L to133.96 mg/L in Shakti Farm (Main Market) and Uttam Nagar, respectively.

8.4.10 CALCIUM CONTENT

The maximum Calcium content in the water sample collected was 26.45 mg/L at Shankar Farm Gutiya minimum 8.02 mg/L at Vaikanthpur (Panch Quarter). It could be because of calcium salts. Water with high calcium content is undesirable to use.

8.4.11 MAGNESIUM CONTENT

The maximum magnesium content was found in 32.68 mg/L at Uttam Nagar and minimum of 7.60 mg/L at Shakti Farm (Main Market). The high magnesium content could be because of the accumulation of magnesium salts and because of this it could have diuretic, catharatic, laxative effects if it is present in high concentrations.

8.4.12 SODIUM CONTENT

The highest Sodium content found 22 mg/L at Sirga, Shakti Farm and minimum of 8 mg/L at Vaikanthpur and Shankar Farm.

8.4.13 POTASSIUM CONTENT

The highest potassium content was found at Shadaura 24 mg/L and minimum of 14 mg/L at Vaikanthpur and Shankar Farm.

8.4.14 CONDUCTANCE

The conductance of the water samples varied from 330 µs/cm at Vaikanthpur (Panch Quarter) to 1520 µs/cm at Uttam Nagar. The value at Uttam Nagar indicates the concentration of total dissolved solids, which is confirmed from the TDS value (1020 mg/L) at the location. The water from five locations has EC value more than 750 µs/cm but less than 2500 µs/cm, indicating water in medium category from the point of irrigation. The water from shallow aquifer of these locations should be used for irrigation with precaution and ample drainage facilities.

8.4.15 AMMONIUM

The ammonia of sample varied from minimum of 0.01 mg/L Shahdaura, Sisona, Shamalpura, Vaikanthpur and Govind Nagar to maximum of 58.15 mg/L Shankar Farm Gautiya.

8.4.16 NITRATE

The nitrate of the water sampled varied from 0.5 mg/L sisaiya to 1.87 mg/L Uttam Nagar. The less value of nitrate shows that there is very less leaching of fertilizer in the aquifer.

8.5 SUMMARY

Survival of living organism and socioeconomic development depend largely on the availability of water. Ground water is a main source of water for irrigation, industries, recreation, domestic and drinking purpose.

With the increasing population and industries, the requirement of water is increasing day by day and as industries are growing the pollutants from these are also increasing and so the quality is decreasing day by day at an alarming rate.

All the samples were collected and analysis in the Water Quality Laboratory of the Department of Irrigation and Drainage Engineering. The samples were analyzed using standard methods. Based on the results of analysis, the following are the conclusions:

a. The ground water samples taken from the shallow aquifers through hand pumps were found odorless and colorless, except at Viakanthpur no.1 and Uttamnagar near Gurudwara Buddha Sahib, were found to have light yellow color.

b. The study revealed that the shallow aquifer ground water was not safe for the drinking purpose with respect to all the parameters taken together. It shouldn't be used for drinking without treatment of particular parameter as the concentration of various constituents was found beyond the safe limits given by BIS.

c. The water samples from most of the locations were found suitable for irrigation purpose as the samples have the concentrations of Total Dissolved Solids, the criteria for measuring EC within the limits prescribed by BIS.

ACKNOWLEDGEMENTS

This study was conducted by the authors under the supervision of Dr. Vinod Kumar, as part of a special project in the Department of Irrigation and Drainage Engineering, College of Technology G.B. Pant University of Agriculture and Technology Pantnagar–263145, Uttam Singh Nagar, Uttarakhand, India. The author express special regards to Dr. Vinod Kumar Professor, Department of Irrigation and Drainage Engineering, Supervisor of this study, for his constant inspiration, invaluable guidance, continuous motivation, deep interest and immense knowledge for bringing out this work.

KEYWORDS

- aquifer
- bis
- deforestation
- ground water
- ground water pollution
- irrigation
- leaching of nitrogen
- living organism
- nitrogen

- socioeconomic development
- total dissolved solids
- tropical zone
- water quality

REFERENCES

1. Bala, I., & Rawat, N. (2006). Study of nitrogen concentration in ground water In Udham Singh Nagar district. Unpublished Dissertation report. Department of Irrigation and Drainage Engineering, College of Technology, G.B. Pant University of Agriculture and Technology, Pantnagar.
2. Chada, D. K., & Tamta S. R. (2000). Ground water pollution in urban areas and its effect on ground water regime, Tenth National Symposium on Hydrology, 428–439.
3. Faillat, J. P., & Rambaud, A. (1991). Deforestation and leaching of nitrogen as nitrate in to underground water in inter tropical zone. Environmental Geology and Water Science, p. 275.
4. Handa, B. K. (1987). "Utilization of saline ground water for irrigation use in semiarid parts of India." Seminar on Groundwater Development, 463–472.
5. Hariharan, A. V. N. S. H. (2007). Some studies on the water quality. parameters of Vuda (Mithilapury) colony Visakhapatnam. Journal of Industrial Pollution Control 23(1), pp. 113–117.
6. Hiremath, S. C., Yadawe, M. S., Puheri, U. S., & Pujara, A. S. (2010). Evaluated physic-chemical parameters of ground water of Bijapur city study conducted was seasonal and used to judge the water quality and suitability for domestic use.
7. Kaushik, N. K. (1963). A study of well in rural Delhi. J. Environ Health. 5, 128–138.
8. Kumar, A. (1994). Ground water pollution with special reference to nitrate in Tarai Bhabhar region of Western U.P. Unpublished MSc thesis. Department of Environmental Sciences. College of Basic Sciences & Humanities, G.B. Pant University of Agriculture and Technology, Pantnagar.
9. Lakhera, S., & Singh, R. (2008). Study of the effect of the monsoon on the water quality of shallow aquifer in Tarai of Udham Singh Nagar district. Unpublished Dissertation report. Department of Irrigation and Drainage Engineering, College of Technology, G.B. Pant University of Agriculture and Technology, Pantnagar.
10. Mahayana, A. C., & Suresh G. C. (1989). Chemical quality of ground water of Manglore city, Karnataka. J. Environ. Health 31(3), pp. 228–236.
11. Nagarnaik, P. B., & Bhalme, S. P. (2012). Study was based on the analysis of drinking water parameter in the education institute situated in Hingna MIDC area, Nagpur.
12. Olaniya, M. S. (1969). Well water quality. Indian J. Environ. Health, 11, 378–391.

13. Pandey, K. S. (2001). A comprehensive pollution study of surface water sediments and ground waters, All India Seminar on Infrastructure Development in Uttarakhand, Problem and Prospects, 4, 89–95.

14. Sahgal, V. K., Sahgal, R. K., & Kaker, Y. P. (1989). In Appropriate Methodology for Development and Management of Ground Water Resources in Developing Countries. 2(3), pp. 879–891.

15. Salade, Y. E., & Sims, J. T. (1992). Evaluation of thiosulfate as a nitrification indication for manures and fertilizers. Plant Soil 147(2), 283–291.

16. Schmidt, K. D. (1974). Nitrate and ground water management development on water quality. J. Amer. Water. pp. 65, 358.

17. Sehgal, M. (2014). Analyzed 30 water sample from Dhakuakhana sub division of Lakhimpur district, Assam, India for various parameters and found out all the parameters were in the permissible limits.

18. Sethi, P. (2013). Analyzed 20 samples for different parameters of drinking water and irrigation water and found that the water of study area was suitable for both drinking as well as irrigation.

19. Sharma, H. C., & Chauhan, H. S. (2001). Continuous reduction in discharge well in Tarai belt of Uttar Pradesh. Challenges in Ground Water Development. 172–175.

20. Sharma, J. D., & Sharma P. (2008). Ground water quality of selected villages of Alwar district, Rajasthan. J. Ecotoxicol. Envion. Mon U. 18(6), 581–586.

21. Shrivastava, R. S., Faego, V. S. Sai and K. C. Mathur, (1998). Water quality along the some river polluted by the orient paper mill water air. Soil Pollution 39(12), 75–80.

22. Singh, A., Nautiyal, A., & Chauhan, N. (2007). Changes in ground water quality of Tarai region in last one decade. Unpublished Dissertation Report. Department of Irrigation and Drainage Engineering, College of Technology, G.B. Pant University of Agriculture and Technology, Pantnagar.

23. Somasundaram, M. V., Ravindaram G., & Mathur, K. C. (1993). Ground water pollution of the Madras urban aquifer. J. India Ground Water, vol. 31, pp. 225–235.

24. Stiger, T. Y., Ribeiro, I., & Carvalho, A. M. M. (2006). Dill application of ground water quality index as an assessment and communication tool in agro environmental policies: Two Portugese case studies. Journal of Hydrology, 327, 578–591.

25. Tase, N. (1992). Ground water contamination. III Japan. Environ. Geological Water Science. pp. 15–20.

APPENDICES

TABLE A-1   Quality Criteria for Designated Best Use of Water

Designated Best-Use	Class of Water	Criteria
Drinking Water without conventional treatment but after disinfection	A	1. Total coliform organism MNP/100 mL shall be or less 2. pH between 6.5 and 8.5 3. Dissolved Oxygen 6 mg/L or more 4. Biochemical Oxygen Demand 5 days 20°C 2 mg/L or less
Outdoor bathing (Organized)	B	1. Total coliform organism MNP/100 mL shall be or less 2. pH between 6.5 and 8.5 3. Dissolved Oxygen 5 mg/L or more 4. Biochemical Oxygen Demand 5 days 20°C 3 mg/L or less
Drinking water source after conventional treatment and disinfection	C	1. Total coliform organism MNP/100 mL shall be 5000 or less 2. pH between 6 and 9 3. Dissolved Oxygen 4 mg/L or more 4. Free Ammonia (as n) 1.2 mg/L or less
Propagation of Wild Life and Fisheries	D	1. pH between 6.0 to 8.5 2. Electrical Conductivity at 25°C μmhos/cm Max. 2250 3. Sodium Absorption Ratio Max. 26 4. Boron max. 2 mg/L
Water with no use for drinking and livestock purpose.	E	Not Meeting A, B, C, & D Criteria 32

TABLE A-2   Indian Standard Drinking Water – Specification (BIS 10500: 1991)

S. No.	Substance or Characteristic	Requirement (Desirable Limit)	Permissible Limit in the absence of alternate source
	Essential Characteristic		
1.	Color (Hazen units, *Max*)	5	25
2.	Odor	Unobjectionable	Unobjectionable
3.	Taste	Agreeable	Agreeable
4.	Turbidity (NTU, *Max*)	5	10
5.	pH Value	6.5 to 8.5	No Relaxation
6.	Total Hardness (as $CaCO_3$) mg/L, *Max*	300	600
7.	Iron (as Fe) mg/Lt, *Max*	0.3	1.0
8.	Chlorides (as CI) mg/L, *min*	250	1000
9.	Residual, free chlorine, mg/L, *min*	0.2	—
	Desirable characteristics		
10.	Dissolved solids mg/L, Max	500	2000
11.	Calcium (as Ca) mg/L, Max	75	200
12.	Copper (as Cu) mg/L, Max	0.05	1.5
13.	Manganese (as Mn) mg/L, Max	0.10	0.3
14.	Sulfate (as S04) mg/L, Max	200	400
15.	Nitarate (as/NO_3) mg/L, Max	45	100
16.	Fluoride (as F) mg/L, Max	1.9	1.5
17.	Phenolic Compounds (as C_6HsOH) mg/L, Max	0.001	0.002
18.	Mercury (as Hg) mg/L, Max	0.001	No relaxation
19.	Cadmiun (as Cd) mg/L, Max	0.01	No relaxation
20.	Selenium (as Se) mg/L, Max	0.01	No relaxation
21.	Arsenic (as As) mg/L, Max	0.05	No relaxation
22.	Cyanide (as CN) mg/L, Max	0.05	No relaxation
23.	Lead (as Pb) mg/L, Max	0.05	No relaxation
24.	Zinc (as Zn) mg/L, Max	5	15
25.	Anionic detergents (as MBAS) mg/L, Max	0.2	1.0
26.	Chromium (as Cr6+) mg/L, Max	0.05	No relaxation

S. No.	Substance or Characteristic	Requirement (Desirable Limit)	Permissible Limit in the absence of alternate source
27.	Poly-nuclear aromatic hydro-carbons (as P AH) mg/L, Max	–	No relaxation
28.	Mineral Oil mg/L, Max	0.01	0.03
29.	Pesticides mg/L, Max	Absent	0.001
30.	Radioactive materials		
	I. Alpha emitters Bq/L, Max	–	0.1
	II. Beta emitters pci/L, Max	–	1.0
31.	Alkalinity mg/L, Max	200	600
32.	Aluminum (as Al) mg/I, Max	0.03	0.2
33.	Boron mg/I, Max	1	5

TABLE A-3   Water Quality for Livestock and Poultry Use

Water Salinity (EC) (dS/m)	Rating	Remarks
< 1.5	Excellent	Usable for all classes of livestock and poultry.
1.5–5.0	Very Satisfactory for Livestock	Usable for all classes of livestock and poultry. May cause temporary diarrhea in livestock not accustomed to such water; watery droppings in poultry.
5.0–8.0	Satisfactory for Livestock	May cause temporary diarrhea or be refused at by animal not accustomed to such water.
	Unfit for Poultry	Often causes watery feces, increased mortally and decreased growth, especially in turkeys.
8.0–11.0	Limited Use for Livestock	Usable with reasonable safety for dairy and beef cattle, sheep, swine and horses. Avoid use for pregnant or lactating animals.
	Unfit for Poultry	Not acceptable for poultry.
11.0–16.0	Very Limited Use	Unfit for poultry and probably unfit for swine. Considerable risk in using for pregnant or lactating cows, horses or sheep, or for the young of these species, In general, use should be avoided although older ruminants, horses, poultry and swine subsist on waters such as these under certain conditions.
>16.0	Not Recommended	Risks with such highly saline water are so great that it cannot be recommended for use under any conditions.

TABLE A-4　Physico-Chemicals Parameters and Their Significance

Parameters	Health Hazards
pH	A measure relative acidity of the water. It is useful in assessing the corrosivity of water to plumbing
Alkalinity	The amount of bicarbonate, the major anion in water, is related to pH and causes corrosion
Acidity	Not a pollutant, in water it neutralizes hydroxyl ions
Solids (dissolved, Suspended and total)	These provide a measure of the suspended solids that can be separated with a filter and the dissolved salts that are present in water.
Conductivity	A measure of total dissolved minerals in water. A change in conductivity or unusual ratio of conductivity to hardness may signal presence of contaminants
Hardness	A measure of the amount of calcium and magnesium. Hardness is a measure of the capacity of water to precipitate soap. This is particularly important if water softening is considered
Chloride	An indicator ion that if found in elevated concentration, pointes to potential contamination form septic systems
Fluoride	Excessive levels of fluorides causes fluorosis, a mottling of the surface of the teeth fertilizer, landfills, or road salt
Nitrate	Nitrate in drinking water can cause blue baby syndrome in infants under six months old. Blue baby syndrome, or methemoglobinemia, are common symptoms of nitrate contamination. Nitrate contamination in drinking water may also increase cancer risk, because nitrate is endogenously reduced to nitrite and subsequent introsation reaction give rise to N-nitrose compounds. These comps are highly carcinogenic and can act systematically.
Calcium	Water with high calcium content is undesirable for various house hold uses, such as washing, bathing and laundering, because of consumption of more soap and other cleaning agents
Magnesium	Mg has diuretic, cathartic and laxative effects if it is present in high concentrations

Table A-5 Guidelines for Interpretation of Water for Irrigation

Potential Irrigation Problem	Units	Degree of Restriction on Use		
		None	Slight to Moderate	Severe
Salinity (affects crop water availability)				
EC$_w$	dS/m	<0.7	0.7 – 3.0	>3.0
(or)				
TDS	mg/L	<450	450 – 2000	>2000
Infiltration				
(affects infiltration rate of water into the soil. Evaluate using EC$_w$ and SAR together)				
SAR = 0–3		>0.7	0.7 – 0.2	<0.2
= 3–6		>1.2	1.2 – 0.3	<0.3
= 6 – 12		>1.9	1.9 – 0.5	<0.5
= 12 – 20		>2.9	2.9 – 1.3	<1.3
= 20 – 40		>5.0	5.0–2.9	<2.9
Specific Ion Toxicity (affects sensitive crops)				
Sodium (Na)				
Surface irrigation	SAR	<3	3–9	>9
Sprinkler irrigation	mg/L	<3	>3	
Chloride (Cl)				
Surface irrigation	mg/L	<4	4 – 10	>10
Sprinkler irrigation	mg/L	<3	>3	
Boron (B)$_s$	mg/L	<0.7	0.7 – 0.3	>3.0
Trace elements				
Miscellaneous Effects (affects susceptible crops)				
Nitrogen (NO$_3$ – N)	mg/L	<5	5 – 30	>30
Bicarbonate (HCO$_3$)	mg/L	<1.5	1.5 – 8.5	>8.5
pH		Normal Range 6.5–8.4		

CHAPTER 9

ELECTROKINETICS APPLICATION TO CONTROL NITRATE MOVEMENT IN SOILS

VINOD KUMAR TRIPATHI

CONTENTS

9.1   ELECTROKINETICS

This chapter discusses application of electrokinetics to control nitrate movement in soils. Electrokinetics is generally defined as physicochemical transport of charge, action of charged particles, and effect of applied electric potentials on formation and fluid transport in porous media. Therefore, the term electrokinetics refers to coupled fluid and charge (electric current) transport.

Electrokinetics has emerged as an innovative method for in-situ restoration of contaminated hazardous waste sites. Direct currents (DC) are applied across electrodes inserted in the soil to generate an electric field for mobilization and extraction of contaminants and for biogeochemical modifications of polluted soils and slurries. Electrokinetic phenomena identified in soils include:

- **Electroosmosis**: If the soil is placed between to electrodes in a fluid, the fluid will move from one side to the other when an electromotive force is applied.
- **Electrophoresis**: it is movement of charged particles (clay particles or microorganisms) suspended in a liquid due to application of an electric potential gradient.
- **Streaming potential**: It is reverse of the electro-osmosis. It defines the generation of an electric potential difference due to fluid flow in soils.
- **Electrolysis**: chemical reactions due to the electric field.

The electrokinetic phenomena that affect electrokinetic remediation are electro-osmosis and electrophoresis [1]. The other two phenomena are not related directly to electrokinetic soil remediation.

9.1.1 VOLTAGE AND CURRENT LEVELS

- Current density and voltage gradients depend on the soil EC.
- Soils with higher EC require more charge and higher currents than lower EC.
- Voltage gradient 100 V/m can be used as an estimate for initial processing.
- Increase current density (voltage gradient) will increase transport rates under ionic migration.
- Other factors like soil properties, electrode spacing and time requirements of the process.

- By Neuromuscular electrical simulation test it is reported that 100 mA is an individual tolerance to electricity.
- Plants and animals can tolerate safely up to 80 mA current between subsurface electrodes.

9.2 NITROGEN TRANSFORMATION IN SOIL

Nitrogen occurs in soil in several forms and inter conversion between these forms is the net result of a large number of dynamic processes (Figure 9.1). Many of these processes are mediated by micro-organisms [6]. While incorporation of ammonium into organic compounds by microbial assimilation is known as immobilization, the reverse process where microorganisms oxidize organic matter to produce energy and convert organic nitrogen into inorganic nitrogen is known as mineralization and both these processes occur simultaneously. In most soils, ammonium is rapidly converted to nitrate via nitrite by a process called nitrification, where ammonium is oxidized to nitrite and then to nitrate by the action of the aerobic bacteria such as *Nitrosomonas, Nitrosospira, Nitrosococcus* or *Nitrosovibrio* and *Nitrobacter, Nitrospina* or *Nitrococcus*, respectively. Ammonium is adsorbed on clay minerals and therefore is less mobile but nitrate is highly mobile. Plants take up nitrogen in mineral form (ammonium or nitrate). Nitrate is very soluble; and unless intercepted and taken up by plant roots leach down in the soil along with irrigation or rain water or it is carried away by runoff. Under some conditions, depending on the availability of organic carbon and anaerobic conditions nitrate may undergo bacterial conversion to molecular nitrogen or nitrous oxide, by a process called 'denitrification.' Unlike nitrifying bacteria, denitrifying bacteria include a wide range of bacteria.

In India crop production has been increased by green revolution. The main factors responsible for success of green revolution are: high yielding crop variety, irrigation and increase in supply of nutrients in form of fertilizer. Rate of consumption of fertilizer is directly proportional to growth rate in production. Among fertilizers, N has played very important role and consumption of N fertilizer has been increased by exponentially (Table 9.1). In India, N fertilizer applications to cereals are increasing faster than

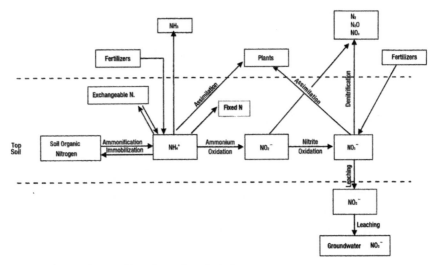

FIGURE 9.1 Nitrogen Transformations in Soil

cereal yields, resulting in declining nutrient use efficiency. This trend can be explained by a fertilizer subsidy regime that has contributed to unbalanced and inefficient fertilizer use.

9.3 MINIMIZING NITRATE LEACHING

Although technologies are available for decontaminating nitrate-containing drinking water, these are very costly. Therefore, it is necessary to minimize nitrate leaching from agricultural as well as non-agricultural activities. In developed countries, the approach to reduce nitrate pollution is largely based on changing the land use pattern. Any change which may involve conversion from agriculture to nonagricultural use, conversion to grasslands, change in cropping system or reduction in fertilizer dose involving a reduction in food production, would not be appropriate in India and other developing countries where the challenge is to maximize food production with no increase in cultivable land. In developing countries, the emphasis will be on maximizing the fertilizer use efficiency (FUE). By adopting the following improved methods of nitrogen fertilizer use, nitrate leaching can be minimized:

TABLE 9.1   Food Production and N Fertilizer Consumption in India

Year	Food production (Mn tons)	N fertilizer Consumption (Mn tons)
1950–51	50.8	0.06
1955–56	68.9	0.11
1960–61	82	0.21
1965–66	72.4	0.57
1970–71	108.4	1.5
1975–76	121	2.15
1980–81	129.6	3.7
1985–86	150.4	5.7
1990–91	176.4	7.99
1995–96	180.4	9.82
2000–01	196.8	10.92
2005–06	208.6	12.72
2006–07	217.28	13.77
2007–08	230.78	14.42
2008–09	234.47	15.09
2009–10	218.11	15.58
2010–11	244.49	16.56
2011–12	259.32	17.30
2012–13	255.36	16.82
2013–14	—-	16.75

Source: Fertilizer Association of India (2011). (Source: http://www.faidelhi.org/general/Consumption%20of%20Fertilisers.pdf)

- Substituting part of the inorganic fertilizers with organic fertilizers, i.e., adopting integrated nutrient management systems.
- Matching the plant needs and fertilizer applications by using appropriate split applications.
- By using improved crop management practices.
- Use of slow-release fertilizers (urea-aldehyde polymeric compounds, coated fertilizers)

TABLE 9.2   N-Use Efficiencies in Different Crops and Locations

Crop	Country	N-use efficiency
Corn	Iowa (USA)	29–45%
(*Zea mays*)	USA	39.6%
Wheat	India	50–55%
(*Triticum aestium* L.)	Pakistan	37–53%
Rice	India	28–34%
	Pakistan	49%
Oat	Spain	37.9%
Cotton	Australia	33%
Spring wheat	Spain	~ 40%
Barley	Finland	60–70%

- Use of nitrification inhibitors and urease inhibitors. Chemicals known to be safe should be used.
- Choosing the right cropping systems.
- Intercepting nitrate by means of trees or other deep rooted, nitrate mining crops (e.g., alfalfa) or by digging ditches.
- Establishment of information systems and monitoring networks.
- Education of the population in general and farmers in particular.

Nitrate originating from agriculture can be a major threat to environment under many situations. It is important that proper assessment of nitrate pollution should be made by extensive analysis of soil and groundwater samples. It is possible to minimize this threat by proper management of agricultural systems and by adopting appropriate policy measures.

9.4   HEALTH ISSUES RELATED TO N POLLUTION IN WATER

Nitrates enter human body through drinking water, food and air. Ingested nitrates are converted to nitrite by microflora and lead to methemoglobinemia, increased free oxide radicals that predispose cells to irreversible damage and effects like cancer, increased infant mortality, abortions, birth defects, recurrent diarrhea, recurrent stomatitis, histopathological changes in cardiac muscles, alveoli of lungs and adrenal glands, deterioration of immune system of the body. Increased consumption of nitrate will lead to [2]:

- Increased production of nitrite;
- Enhanced absorption of sodium from the intestinal lumen;
- Excess NO_X (free radical nitric oxide) generation having vasodilatory effect; and
- Increased production of O^{-2}, which will react with other cell constituents, possibly causing irreversible cell damage.

About 20% of ingested nitrate is reduced to nitrite by nitrate-reducing microflora present in the saliva at the base of the tongue. The factors, which influence oral microflora and hence reduction of nitrate are: nutritional status, infection, environmental temperature and age (more in elderly). Ingested nitrate is reduced to nitrite by nitrate-reducing microflora in the stomach (under favorable conditions, viz. pH $\geq$ 4) and upper part of the intestine. Conditions favoring high stomach pH are achlorhydria, atrophic gastritis, artificially fed infants, or patients using antacid or similar drugs, e.g. Omeprazole.

The nitrite is readily and completely absorbed from both the stomach and the upper small intestine. The absorbed nitrite is then rapidly distributed throughout the tissues. It is rapidly oxidized to nitrate in the blood, with the formation of methemoglobin.

Naturally occurring nitrate levels in surface and groundwater are generally a few milligrams per liter. Higher nitrate levels are found in groundwater due to water percolating through nitrate-rich rocks and also due to an excessive use of chemical fertilizers. The WHO report of 2004 maintains that extensive epidemiological data support limiting the value of nitrate-nitrogen to 10 mg/L or as nitrate to 50 mg/L for human consumption, whereas IS-10500 prescribes maximum permissible limits in drinking water as 45 mg of NO_3/L.

9.5 SOURCES OF NITRATE

The main sources contributing to nitrate content of natural waters are atmosphere, geological features, anthropogenic sources, atmospheric nitrogen fixation and soil nitrogen. Oxides of nitrogen are generated through lightning and reach the surface water with rain. Recent reports indicate that atmospheric contributions amount to 25% of the total load of nitrate. Direct discharge from septic tanks, sewage and industrial effluent are other contributors. Excessive use of chemical fertilizers is one of the main sources of nitrate in water. Out of the total human nitrate intake, fruit and vegetables account for 70%, drinking water 21%, and meat and meat products 6%. The common nitrate-rich vegetables are lettuce, spinach, beetroot, celery, egg plant, beet, banana, strawberry, tomatoes and peas. Preservatives used in the food industry are significant nitrate sources. Cooking in aluminum utensils enhances reduction of nitrates to nitrite, and hence increases the toxicity.

9.6   STATUS OF NITROGEN IN GROUNDWATER

Excessive irrigation in addition to increased intensity of N fertilizer use have resulted in nitrate ($NO3-$) pollution of groundwater and the survey showed that large areas in Uttar Pradesh (300–694 ppm), Punjab (362–567 ppm) and Haryana (300–1310 ppm) have been affected by $NO3-$ pollution of groundwater. The problem of high nitrates in groundwater is both extensive and intensive, posing serious health hazard. However, as irrigation is one of the essential components of the modern agricultural production system, groundwater is utilized to irrigate ~36.25 Mha in India. As a result, N in the form of $NO3-$ (0.11 Tg N) from groundwater is brought into the agricultural production system

Table 9.2 suggests that the story of N in groundwater is more complex, since levels are high compared to the surface environment. This is an explosive issue similar to that of arsenic and fluoride in groundwater in South Asia [8]. On the other hand, the fate of N in river sediments is closely linked to that of the organic carbon, as observed by several studies.

9.7   SOIL SURFACE NITROGEN BALANCE

The soil surface nitrogen balance is calculated as the difference between the total quantity of nitrogen inputs entering the soil and the quantity of nitrogen outputs leaving the soil. Total inputs and outputs along with the

TABLE 9.2   Nitrate in Sub-Surface Waters in Selected Regions in India

Source	Region/District/State	Groundwater (agricultural belt) (NO_3-N mg/L)
Bijay Singh et al.	Agra, UP	54
	Aligarh, UP	61
	Meerat, UP	157
Subramanian	Gujarat	460
	Rajasthan	262
	West Bengal	48
Kanwar Singh et al.	Kanpur (UP)-shallow wells	8–96
Bijay Singh et al.	Mahendragarh (H)	296
	Ambala (H), Hisar (H)	223
	Bathinda (P)	95
Chadha and Sarin	Faridkot (P)	11–500
	Patiala (P)	5–456
	Sangrur (P)	10–900
	Bhiwani (H)	5–480
	Khurukshetra (H)	5–360
	Sonepat (H)	10–1310
Shirahati et al.	Ludhiana (P)	12–30
Kanwar Singh et al. (deep wells)	Industrial (UP)	5–1559
Jack and Sharma	Coimbatore (TN)	9–219
Subramanian	India average	65

soil surface N balance for the entire agricultural land of India as a whole are given in Table 9.3. Nearly 14.6 Tg of nitrogen has been estimated as input from different sources, with output nitrogen of about 12.24– 12.71 Tg. Soil-surface N balance estimated as input minus output is found to be about 1.89–2.32 Tg. Inorganic N fertilizer constituted the major percentage of total inputs, and fodder and feed the major percentage of total outputs estimated the spatial patterns in nitrogen balance for different states of India. It varied from deficit to surplus. The highest nitrogen surplus was

TABLE 9.3   Soil-Surface Nitrogen Balance (Tg) for Agricultural Land of India [9]

N source	Tg N
Inputs	
Inorganic N fertilizer	10.8
Biological N fixation	1.14–1.18
Compost	0.17
Animal waste (manure)	1.53
Wet deposition	0.81
Groundwater	0.11
Total	**14.56–14.6**
Outputs	
Harvested crop	4.13
Fodder	5.81
Fuel	1.9
Erosion loss	0.06
GHG emission	0.34–0.81
Total	**12.24–12.71**
Balance leaching	2.32–1.89

found in Uttar Pradesh (2.5 Tg) and negative nitrogen balance was found in Orissa (–0.01 Tg) and the Northeastern states.

9.8   RESEARCH ON ELECTROKINETICS: CASE STUDY

The study was conducted at Department of Agricultural and Biological Engineering, University of Florida, USA with the objective to determine whether nitrate movement can be effectively controlled by electrokinetics in a vertical soil column.

The effectiveness of controlling nitrate movement with electrokinetics was evaluated in soil column laboratory experiments. The experimental setup and procedures were intended to model drip irrigation methods with future field crop applications envisioned. In arid climates, drip irrigation, which applies water directly in the vicinity of the root zone, is believed to be one of the most efficient irrigation methods, wetting a limited amount of surface area and depth of soil. A major reason that drip irrigation was considered in the study was that electrical wires can be simultaneously buried by drip tube insertion equipment [3, 4].

9.8.1 EXPERIMENTAL PROCEDURE

a.   Prepare the soil slurry by mixing nitrate-free tap water with sandy soil,
b.   Prepare the nitrate solution and store it in a tank,
c.   Pack the column and insert electrodes, tensiometers, and wires,
d.   Flush the soil column with tap water,
e.   Apply sodium nitrate ($NaNO_3$) solution,
f.   Allow the electric current, and
g.   Take soil samples from different soil layers at desired times.

9.8.1.1   Soil Column

An acrylic column, 38 cm in height by 14 cm in diameter, was used in all experiments. Two electrodes, spaced 25 cm apart, were placed in the column, the anode at the top of the column and the cathode at the bottom. Electrodes were encased in soil by placing thin layers of soil below the cathode and atop the anode, and the electrodes were connected to an external, adjustable, constant current electrical power supply. The drainage port permitted effluent drainage to an open container.

Sampling ports were spaced 5 cm apart, the top ports located 2.5 cm below the anode, and the bottom (fifth) ports 2.5 cm above the cathode. Four 1.3 cm diameter-sampling holes were drilled through column walls at each layer. A 0.7 cm diameter sampling tube was used to take soil samples from the edge to the column interior; yielding a fan-shaped sampling area. Tensiometers were inserted into each layer for soil water status measurement, and the results were recorded with a data logger. A voltage divider, a device to decrease or increase voltage proportionally, was connected between the wires and the data logger to ensure the recording of the voltage in the data logger range (± 5 V).

9.8.2   NITRATE MOVEMENT AS A FUNCTION OF CURRENT DENSITY

Constant current inputs of 60 mA and 100 mA were also applied to the soil columns that had been loaded with solute containing 221 mg/L of nitrate. Figure 9.2 shows that the higher the electrical input was, the larger the resulting nitrate concentration near the anode. After equal current application periods (6 h), the 100 mA current input resulted in 1048 mg/L of NO_3^- in the layer 2.5 cm from the anode, while the concentrations in this layer were 327 and 445 mg/L with 60 mA and 80 mA, respectively. After 6 h, the nitrate concentration in the system with no electrical input is almost uniformly distributed. However, the highest nitrate concentration was observed 7.5 cm from the anode. This suggests that this was the driest layer. The water content at the 7.5 cm layer was 14.61%, but at the bottom layer (22.5 cm) the water content was 21.77%. It is theorized that the placement of the electrodes may have disturbed the water flow, and thus the nitrate distribution [7].

9.8.2.1   Change of pH

The pH changes were also measured immediately after each sampling using pH paper. Small pH changes were observed in the soil samples taken between the 2.5 cm and 22.5 cm layers. Because very high NO_3^- levels were measured adjacent to the anode, pH was also determined close to the electrodes. The results are shown in Figure 9.3. After 6 h electricity at 80 mA, the pH was 3.5 near the node and 11.0 near the cathode, indicating that electrolysis occurred at the electrodes. The pH values found in the column at intermediate sampling locations were relatively constant and similar, suggesting that the soil has some buffering properties.

9.9   CONCLUSIONS

a. Electrokinetics is the effective tool when treating heterogeneous, silt and clay rich soil.
b. Electroosmosis gives dual benefits of uniform soil heating and uniform pore water movement.

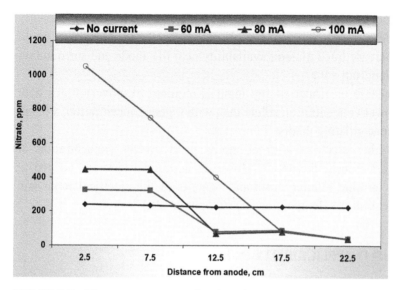

FIGURE 9.2   Nitrate movement as a function of current.

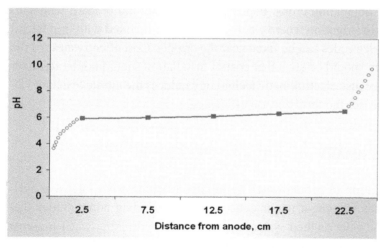

FIGURE 9.3   The pH gradient due to 6 h of electrical input at 80 mA.

   c. An electrical input can increase nitrate content adjacent to the anode in sandy soil to an extremely high nitrate concentration level.

   d. The electrical input will consequently reduce the nitrate concentration near the cathode to a very low level, thus minimizing loss from that location.

e. An electrical input will reduce the pH near the anode significantly and similarly increase the pH near the cathode. This effect could increase trace mineral availability near the anode and surrounding plant root zone.

f. A large electrical energy input is required to move nitrates with lower concentration solute than with higher concentration, a result substantiating theory.

g. Higher electrical current inputs yield greater concentration of nitrates near the anode. However, additional tests are needed to determine relative costs and benefits of the electrical input and determine safe upper electrical input levels.

9.10   FIELD APPLICABILITY IN INDIA

The technology is conceptually correct and under research process. But for the application point of view in Indian condition it is feasible only in low plant population crops. Especially in horticultural orchard with sub-surface drip irrigation where electrical wires can be fixed with drip lateral pipe and electrodes can be fixed near the dripper. Cost effectiveness of the technology should be such that cost of installation should not be fertilizer saving. Because electrokinetic technology reduces the nitrate leaching vis-à-vis increases the fertilizer use efficiency.

9.11   SUMMARY

Contamination of groundwater by nitrates leaching from intensive agricultural has become a major concern for surrounding communities that use groundwater as their water supply. Nitrate in drinking water presents a human health risk, i.e., methemoglobinaemia ("blue baby" syndrome) [5]. The major source of nitrate contamination is believed to be nitrogen fertilizer from agricultural fields.

Best Management Practices have been developed to guide fertilizer use and minimize nitrogen losses, but they do not address control of nitrate movement from the crop root zone. Electrokinetics has emerged as an innovative method for in-situ control of nitrate movement, retaining it near the root zone. Direct currents (DC) are applied across electrodes inserted in the soil to generate an electric field for mobilization of nitrate in soils. Study is on nitrate movement and pH changes in a vertical, partially saturated sandy soil column subjected to an electrical current. The highest measured nitrate concentration (7155 ppm) was within 5 mm of the anode after application of a 6 h 80 mA current. The nitrate concentration at the cathode was 1/5th of the inflow solute concentration. The pH was 11 near the cathode, 3.5 near the anode, and showed little change in intermediate layers. Studies of different nitrate concentrations and electrical current levels suggest that 80 mA electrical current applied for 6 h duration might effectively control nitrate migration in similar sandy soil conditions.

KEYWORDS

- anode
- cathode
- current
- electrokinetics
- electroosmosis
- electrophoresis
- food production
- groundwater
- health
- methemoglobinemia
- nitrate
- nitrogen
- nitrogen consumption
- nutrient use efficiency
- pollution
- soil
- subsurface water
- transformation
- voltage

REFERENCES

1. Acar, Y. B., Gale, R. J., Alshawabkeh, A. N., Marks, R. E., Puppala, S., Bricka, M., & Parker, R. (1995). Electrokinetic remediation: Basics and technology status. J. Hazard. Mater., 40(2), 117–137.
2. Gupta, S. K., Gupta, R. C., Chhabra, S. K., Sevgi Eskiocak, Gupta, A. B., & Rita Gupta (2008). Health issues related to N pollution in water and air. Current Science, 94(11), 1469–1477.
3. Jia, X., Larson, D. L., & Zimmt, W. S. (2005). Effective nitrate control with electrokinetics in sandy soil. Trans. ASABE, 49(3), 803–809.
4. Jia, Xinhuav Dennis, L., Donald, S., & James W. (2005). Electrokinetic control of nitrate movement in soil. Engineering Geology, 77, 273–283.
5. Majumdar, D. (2003) The Blue Baby Syndrome: Nitrate Poisoning in Humans. Resonance, Oct., pp. 20–30.
6. Prakasa Rao, E. V. S., & Puttanna, K. (2000). Nitrates, agriculture and environment. Current Science, 79(9), 1163–1168.
7. Rawahy, S., Larson, D. L., Walworth, J., Slack, D. C. (2003). Effect of electrical input with drip irrigation on nitrate distribution in soil. Applied Eng. in Agric. (ASABE), 19(1), 55–58.
8. Subramanian, V. (2008). Nitrogen transport by rivers of south Asia. Current Science, 94(11), 1413–1418.
9. Velmurugan, A., Dadhwal, V. K., Abrol, Y. P. (2008). Regional nitrogen cycle: An Indian perspective. Current Science, 94(11), 1455–1468.

PART 4

SCOPE OF EMERGING AGRICULTURAL CROPS

CHAPTER 10

SCOPE OF MEDICINAL AND AROMATIC PLANTS FARMING IN EASTERN INDIA

K. M. SINGH

CONTENTS

Paper presented in the "Trainers Training Program. Under the scheme *Scaling up of Water Productivity in Agriculture for Livelihoods through Teaching-cum-Demonstration*, organized from 19 February to 4 March 2009, at ICAR Research Complex for Eastern Region. Patna-14. Electronic copy available at: http://ssrn.com/abstract=2019789.

10.1 INTRODUCTION

The use of plants as medicine is older than recorded history. As mute witness to this fact marshmallow root, hyacinth, and yarrow have been found carefully tucked around the bones of a Stone Age man in Iraq. In 2735 B.C., the Chinese emperor Shen Nong wrote an authoritative treatise on herbs that is still in use today. Shen Nong recommended the use of Ma Huang (known as ephedra in the Western world), for example, against respiratory distress. Ephedrine, extracted from ephedra, is widely used as a decongestant. We find it in its synthetic form, pseudoephedrine, in many allergy, sinus, and cold-relief medications produced by large pharmaceutical companies.

10.1.1 MEDICINAL PLANTS AND HUMAN HEALTH

South Asia is home to many rich, traditional systems of medicine (TSM). Ayurvedic system dates back to 5000 B.C. Along with the Unani, Siddha and Tibetan systems, these TSMs remain important source of everyday health and livelihood for tens of millions of people. Himalayan sage (scholars of Traditional Medicine) have said "*Nanaushadhi Bhootam Jagat Kinchit, i.e., 'there is no plant in the world, which does not have medicinal properties.'*" The ancient scholars are estimated to know the medicinal properties of hundreds of species of plants. It is therefore, no exaggeration to say that the uses of plants for human health are probably as old as human beings themselves. Even so, the recent dramatic increase in sales of herbal products in global markets underscores the growing popularity of herbal therapies [21].

While this has created new opportunities for the countries, their largely impoverished populace and traditional herbal industry, it also poses unprecedented threats to the very resources on which the industry is dependent besides creating socioeconomic imbalances and erosion of spiritual and cultural heritage and knowledge systems. Medicinal and aromatic plants (MAPs), including trees, shrubs, grasses and vines, are a central resource for these traditional health systems, as well as for pharmaceutical (or allopathic) medicines. There are more than 8,000 plant species in South Asia with known medicinal uses [14].

Medicinal plants are accessible, affordable and culturally appropriate sources of primary health care for more than 80% of Asia's population (WHO). Poor and marginalized, who cannot afford or access formal health care systems, are especially dependent on these *culturally familiar, technically simple, financially affordable* and *generally effective* traditional medicines. As such, there is widespread interest in promoting traditional health systems to meet primary health care needs. This is especially true in South Asia, as prices of modern medicines spiral and governments find it increasingly difficult to meet the cost of pharmaceutical-based health care [17].

Medicinal plants form the basis of traditional or indigenous systems of health used by the majority of the population of most developing countries. In China, India and many other countries in South and East Asia, traditional systems of medicine use thousands of plant species to treat malaria, stomach ulcers, and various other disorders (Table 10.1). Throughout the region, there is strong and sustained public support for the protection and promotion of the cultural and spiritual values of traditional medicines.

10.1.2 HERBAL MEDICINE TODAY

The World Health Organization (WHO) estimates that 4 billion people, 80% of the world population, presently use herbal medicine for some aspect of primary health care. Herbal medicine is a major component in all indigenous peoples' traditional medicine and a common element in Ayurvedic, homeopathic, naturopathic, traditional oriental, and Native American Indian medicine. WHO notes that of 119 plant-derived pharmaceutical medicines, about 74% are used in modern medicine in ways that

TABLE 10.1 Distribution of Medicinal Plants [2, 17]

Country or region	Total number of native species in flora	No. of medicinal plant species reported	% of medicinal plants
World	297,000	52,885	10
India	17,000	7,500	44
Indian Himalayas	8,000	1,748	22

TABLE 10.2   Status of Various Herbal Based Medical Systems in India [2, 17]

Characteristics	Medical Systems				
	Ayurveda	Siddha	Unani	Tibetan	Homeopathy
Dispensaries	15193	444	1193	-	5634
Hospitals	753	276	74	-	223
Licensed pharmacies	8533	384	462	-	613
Medicinal plants known	2000	1121	751	337	482
Post graduate college	57	3	8	-	31
Registered practitioners	438721	17560	43578	-	217460
Under graduate college	219	6	37	-	178

correlated directly with their traditional uses as plant medicines by native culture.

Major pharmaceutical companies are currently conducting extensive research on plant materials gathered from the rain forests and other places for their potential medicinal value (Table 10.2). In all the countries of South Asia, medicinal and aromatic plants (MAPs) play a significant role in the subsistence economy of the people, especially those living in the rural interiors. The collection, simple processing and trading of medicinal plants contribute significantly to the cash income of the poor and women in these regions. A recent study carried out by CECI-India [19], indicated that from a single district of Pithoragarh in Uttranchal state of India, more than 1300 tons of MAPs are collected and traded annually, most of them illegally [17].

Unsustainable and large scale harvesting of MAPs from the natural habitats without providing equitable benefit to the local people and government is of grave concern to all. Therefore, by sustainably using and growing economically remunerative MAPs, there is an ample scope to maintain both the rural livelihoods and environmental sustainability. MAP-based local micro- enterprises can also bridge the gap between rural poor and relatively well-off urban rich and promote social harmonization and sound environment conservation [17].

10.2 CONSTRAINTS AND OPPORTUNITIES IN CULTIVATION

Some of the practical applications integrating medicinal plants into traditional farming systems have taken an obligate relationship in backstopping upland agriculture or mixed farming. South Asian states have a rich and diverse traditions of practicing complex and rotational farming systems that includes herbal plants cultivation [2] and therefore, conservation and *ex-situ* cultivation of medicinal plants especially applying organic farming protocols has a great scope especially to access international markets [12].

Other important opportunity and advantage of cultivating MAPs include ease of their incorporation in the existing cropping systems due to availability of a large number of species and choice of plant types, i.e., trees, shrubs, forbs, vines and their suitability to grown in different eco-physical conditions. To sum up, wild stock management and cultivation of carefully selected species as a mixed, inter or companion crop in agro and farm forestry conditions is feasible and needs to be pursued. However, in order to ensure a good input and service delivery system including marketing cultivation may need to be carried out in selected pockets in an intensive manner [14].

10.3   DEFINITION OF A MEDICINAL PLANT

A considerable number of definitions have been proposed for the term 'medicinal plant.' According to the WHO, "a medicinal plant is any plant which, in one or more of its organs, contains substances that can be used for therapeutic purposes, or which are precursors for chemo-pharmaceutical semi synthesis."

This definition allows for a distinction between the already known medicinal plants whose therapeutic properties or character as a precursor of certain molecules have been established scientifically, and other plants used in traditional medicines and regarded medicinal, but which have not yet been subjected to thorough scientific study.

10.4   PLANTS FOR HEALTH

Today's healthcare systems rely largely on plant material. Much of the world's population depends on traditional medicine to meet daily health requirements, especially within developing countries. Use of plant-based remedies is also widespread in many industrialized countries and numerous pharmaceuticals are based on or derived from plant compounds. Similarly, cosmetics and other household products may contain plants of medicinal or therapeutic value.

The pharmaceutical industry is both large and highly successful. Sales of plant derived drugs were $30 billion worldwide in 2002. At present about 50% of the total plant-derived drug sales come from single entities, while the remaining 50% come from herbal remedies. Although the latter have greater volumes of consumption, the relatively low volumes of single entities, which are mostly prescription products, are more than compensated by their higher prices. Single entity plant drugs, which mostly treat serious medical ills, include atropine, digoxin, morphine, paclitaxel, pilocarpine, reserpine, scopolamine, topotecan and vincristine, among many others. Several of the compounds have outlived their usefulness in light of better alternatives, however, these are exhibiting decline in sales. On the other hand, as a consequence of new drug developments, single entities overall are projected to increase their market share of the combined total future dollar sales.

10.5   THE EASTERN REGION OF INDIA

The Eastern Region of India comprising of eastern UP, Bihar, Jharkhand, West Bengal, Assam, Odissa, and Chhattisgarh has a large number of valuable medicinal plants naturally growing mostly in fragile ecosystems that are predominantly inhabited by rural poor and indigenous communities. The sustainable management of these traditionally used plants not only help conserve nationally and globally important biodiversity but also provide critical resources to sustain livelihoods. In North Bihar plains and forests, for example, we find in abundance, a diverse range of herbs, shrubs, trees and vines that have significant medicinal value whose healing properties are known to the local healers and traditional doctors for centuries but are currently threatened due to lack of concerted conservation efforts, e.g. value of medicinal raw materials annually exported from Nepal to India and other countries is estimated between 18 to 20 million dollar [10]. The Eastern Region can also benefit from the recent interest of the Western world in MAP grown in these areas bordering Nepal and earn a lot more than what Nepal earns by taking advantage of the liberal trade regime which has opened newer markets for our products post – WTO and also the renewed interest of the farmers towards this very lucrative sector [9].

The eastern region of the country, recognized as a 'low productivity high potential' region needs holistic management of land, water, crops, biomass, horticultural, livestock, fishery and human resources. The region, though endowed with rich natural resources has lacked in capitalizing on these resources. Though, it has a rich resource base for intensive and diversified agriculture, still, declining per capita land and water availability, reduced diversion of water for agriculture to meet the food requirement of burgeoning population coupled with degrading land and water resources are a serious challenge towards improving productivity and sustainability in this Region [5].

There is an urgent need to identify suitable crops and develop high yielding, disease and pest resistant, location-specific varieties of important, cereals, oilseed, pulses, fruits, vegetables and medicinal and aromatic plants to meet the global competition [ICAR-RCER, Vision 2025].

10.5.1 GENERAL FEATURES OF THE EASTERN REGION

The eastern region comprising of eastern UP, Bihar, Jharkhand, West Bengal, Assam, Orissa, and Chhattisgarh occupies about 22.5% of the country's geographical area and is inhabited by about 35% of the country's population (Figure 10.1). The region can be divided into 3 distinct physiographical units, namely (i) Plains of eastern UP, Bihar, West Bengal, and Assam; (ii) Hilly and plateau regions in eastern UP, Bihar, Jharkhand, West Bengal, Orissa, Chhattisgarh, and Assam; and (iii) Coastal plains of West Bengal and Orissa.

The climate of the region is tropical, hot and humid except in hilly areas with high rainfall. The average annual rainfall varies from 1025–2820 mm. Even though the region has rich rain, surface and ground water resources, they are grossly underutilized, with the result that a large proportion of the cultivated area does not receive any irrigation water. The farmers depend on the vagaries of the monsoon for crop production.

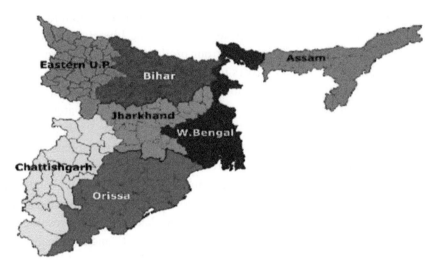

FIGURE 10.1   Spread of Eastern Region in India [4, 5].

Owing to poor utilization of water resources, the cropping intensity in the region is low. Perennial and seasonal water bodies abound in this region but their potential has not been exploited. These provide a great scope for development of medicinal crops like sarpagandha, buch, shatawari, musli, etc. and aromatic crops like mint (mentha), lemongrass, java citronella, palmarosa, jama rosa, tulsi, etc. The eastern region has specific advantages as well as handicaps [5].

Though, eastern region is endowed with natural resources but, so far its potential could not be harnessed in terms of improving agricultural productivity, poverty alleviation and livelihood improvement.

10.5.2 SOCIAL PERSPECTIVE: EASTERN REGION OF INDIA VERSUS HERBAL WAY

Use of medicinal plants in primary health care and nutrition needs is traditional and imbedded in all cultures. No major problems of acceptability regarding familiarity with the usage of plant products, methods of cultivation of many commonly grown plants and technologies required for processing into items of common household uses and value. Med-plants have also been used to develop family- based health and livelihood oriented enterprises in rural areas. MAPs help to: a) Preserve the traditional medical knowledge, b) Provide easily adaptable enterprising opportunities for unemployed youth and rural poor who can learn the trade from their parents and peers and earn not only their livelihood but also contribute to the society [10].

10.5.3 PROTECTION OF TRADITIONAL KNOWLEDGE: EASTERN REGION OF INDIA VERSUS HERBAL WAY

The urgency and need to protect the fast disappearing medicinal plants-based traditional knowledge, which is still abundant in Eastern Region, cannot be overemphasized. If proper values can be added to the traditional medical knowledge-based health practices and subsistence-oriented MAP applications, a large number of jobs can be created in the rural areas. Even at current level of conversion of traditional medicinal knowledge into economic opportunities, enterprise-based application can account for thousands of jobs in rural areas of the Eastern Region. Thus, med-plants have high potential in creating jobs and pushing economic growth in resource-constrained areas suffering from limited educational opportunities, lack of infrastructure, and underdeveloped med- plants-based trade and commercial activities. The conversion of socio-cultural traditions and indigenous knowledge into livelihood means and economic opportunities also has the advantage of preserving the rapidly eroding cultural knowledge and practices which are increasingly threatened due to globalization and homogenization of people and communities [11].

10.5.4 ENVIRONMENTAL PERSPECTIVE: EASTERN REGION OF INDIA VERSUS HERBAL WAY

The growing apathy toward products made from chemical products becoming ethically unacceptable. This has created new markets for quality, certified and organic herbal products. Medicinal plants have the potential to fill these needs as they provide green health alternatives and a number of other eco-friendly products of domestic and industrial usage. Found as trees, shrubs, grasses and vines, these plant species abundantly growing in the plains of eastern region. Its entry into the world food and drug market as the environment friendly botanical products is looked upon as an emerging and new opportunity. The development of medicinal plants-based economic incentives is being increasingly applied to enlist greater participation of people in conservation of forest ecosystem.

10.5.5 POTENTIAL OF EQUITABLE COMMERCIALIZATION: EASTERN REGION OF INDIA VERSUS HERBAL WAY

The MAP sub-sector has immense potential as the sustainable commercialization can benefit farmers and industry both by providing higher price and by opening up national and global markets for new products from the region. Private sectors stand to benefit by ensuring sustainable supply of quality raw materials to benefit their industry and trade if they can be facilitated to build partnerships with farmers. Many of the species are shade tolerant and others are climbers, trees, shrubs and herbs that can be grown in different configurations of crop geometry.

10.5.6 TRADE AND ENTERPRISE DEVELOPMENT: EASTERN REGION OF INDIA VERSUS HERBAL WAY

The demand for medicinal and aromatic plants in India—to meet both domestic and export market—comprising 162 species, increased from 15 to 16% between 2002 and 2005 [2]. Medicinal and aromatic plants cultivation and management therefore, can become highly remunerative both in financial and economic terms for the small-scale growers.

The current gap between demand and supply is estimated to be 40,000 to 200,000 tons, which increased from 152,000 to 400,000 tons in 2009. Not only the plants are in increasing demand by major herbal drug industries as an essential raw material of their drugs, but also its collection, production, processing, packaging and transportation requires high labor input, which can create employment in job-starved eastern region of India. Collection from wild and selective harvesting in addition to primary processing is mostly done manually, and even at the secondary and tertiary levels, med-plants have substantial labor requirements. Moreover, not only do MAP-based industries expand jobs, enhancing traditional uses through value added processing can increase cash earnings to the local people [12].

10.5.7 WORLD MARKET TRENDS: EASTERN REGION OF INDIA VERSUS HERBAL WAY

The European market for herbal supplements is estimated at over US$ 2.7 billion and for herbal remedies, a further US$ 0.9 billion. Germany is by far the largest market. The market is growing rapidly at over 4% per annum for herbal remedies and considerably faster for herbal supplements. The US herbal market is nearing saturation and is expected to peak at US$ 6 to 8 billion in the next few years. Demand for medicinal plants is expected to continue to expand rapidly, fueled by the growth of sales of herbal supplements and remedies. Their basic uses in medicine will continue in the future, as a source of therapeutic agents, and as raw material base for the extraction of semi- synthetic chemical compound such as cosmetics, perfumes and food industries [2].

10.5.8 DOMESTICATION AND CULTIVATION OF MAP: EASTERN REGION OF INDIA VERSUS HERBAL WAY

Some of the practical applications integrating medicinal and aromatic plants into traditional farming systems have taken an obligate relationship in backstopping upland agriculture. Other important opportunities and advantages of cultivating MAPs include ease of their incorporation in the existing cropping systems due to availability of a large number of species and choice of plant types. Cultivation of carefully selected species as a mixed, inter or companion crop in agro and farm forestry conditions following a soil-improving crop rotation is highly feasible livelihood enhancing activities in Eastern region. However, this will require an improved input and service delivery system including marketing, and postharvest technologies. Cultivation needs to be done on a business platform by a chain of small and micro-enterprise-based groups and individuals. In order to achieve an economy of scale and desired impact, it may need to be concentrated in selected pockets in an intensive manner as cluster of activities and micro-enterprises.

10.6   CULTIVATION OF MEDICINAL PLANTS

Information on the propagation of medicinal plants is available for less than 10% and agro- technology is available only for 1% of the total known plants globally. This trend shows that developing agro-technology should be one of the thrust areas for research. Furthermore, in order to meet the escalating demand of medicinal plants, farming of these plant species is imperative. Apart from meeting the present demand, farming may conserve the wild genetic diversity of medicinal plants. Farming permits the production of uniform material, from which standardized products can be consistently obtained [17].

Cultivation also permits better species identification, improved quality control, and increased prospects for genetic improvements. Selection of planting material for large-scale farming is also an important task. The planting material therefore should be of good quality, rich in active ingredients, pest- and disease-resistant and environmental tolerant. For the large scale farming, one has to find out whether monoculture is the right way to cultivate all medicinal plants or one has to promote poly-culture model for better production of medicinal plants.

Studies conducted on the agro-forestry of medicinal plants elsewhere suggest that since many medicinal plant species prefer to grow under forest cover, agro-forestry offers a convenient strategy for their cultivation as well as conservation through:

1.   Integrating shade tolerant medicinal plants as lower strata species in multi-strata system;
2.   Cultivating short cycle medicinal plants as intercrops in existing stands of tree crops;
3.   Growing medicinal tree as shade providers and boundary markers; and
4.   Inter-planting medicinal plants with food crops.

Notwithstanding, it is understood that the cultivation of medicinal plants is not an easy task as the history of medicinal plants farming reflects.

Many farmers in trans-Himalayan region of northern India have replaced the medicinal plants farming with common crops [i.e., peas (*Pisum sativum*), and potatoes (*Solanum tuberosum*)] due to the lengthy cultivation cycle of medicinal plants. The cost of many medicinal plants in northern India is lower than many seasonal vegetables, which is a cause of scanty farming of medicinal plants. Attempts should however, be made by different organizations to cultivate various medicinal plant species, including rare and endangered categories.

Agro-technology for about 20 species of rare and endangered medicinal plants of the northern India has been developed by different organizations. However, the per hectare cost of cultivation, total annual production and cost benefit ratio fluctuate with different medicinal plant species. At present, however, the farming of most of the medicinal plant species is being operated on a very small scale and is restricted to a few hectares of land in various states of eastern India. There is also a market uncertainty, fear about lower prices and lack of knowledge about the necessary pre and postharvest processes, and needs for obtaining permits from government agencies for cultivation of medicinal plants.

Additionally, many farmers are unaware about the agency responsible for issuing permits. If the farmers are not granted permits needed to cultivate, they are forced to sell their products on the illegal market, which exposes them to action by government agencies and the exploitation by middlemen. The irrigation and harvesting operations have not been mechanized for MAPs. Agricultural Engineers can play pivotal role in the innovative methods of cultivation, processing and preservation of MAPs.

10.7 OPPORTUNITIES IN DEVELOPING THE MEDICINAL PLANT SECTOR

For developing the 'herbal industries,' the eastern India possesses a rich diversity of medicinal plant species across the various forest types along an altitudinal gradient (as discussed in the use and diversity of medicinal plants). Such a high diversity of medicinal plants would be helpful for further scientific research on exploring their medical efficacy, value addition, and use in curing various old and new diseases.

India has already established a reputation as a low-cost manufacturer of high quality generic drugs in the global market. This fact can be used as an important tool for the marketing of herbal products produced in India. It is expected that India's aim to build a golden triangle between traditional medicine, modern medicine, and modern science will be a boon for developing the traditional herbal medicine and the medicinal plants sector [17].

10.7.1 EXISTING POLICIES

a. The way forward

The Forest Departments of the eastern Indian states will have to identify two major areas in each Forest Division; namely the conservation area and the developmental area. The conservation areas will be selected based on their rich medicinal plants diversity and marked for *in-situ* conservation and complete protection in the concerned Forest Division. In the developmental areas, apart from protection of the existing bio-resources, the medicinal plant species of the neighboring areas will also be introduced and cultivated at a large scale. The remaining areas in the Forest Division will remain open for sustainable harvesting of the medicinal plant species. A Joint Harvesting Team, composed of medicinal plants experts, Forest Department officials and some selected local people, will be constituted, which will decide the extent of annual harvesting of the desired medicinal plant species.

The various policies at national and state level and their subsequent implementation will provide an opportunity in the advancement of medicinal plants sector. This model of conservation and cultivation of the medicinal plants may be useful for generating the raw material for the 'Herbal Industries' as well as for ensuring the conservation of the rare medicinal plants.

b. Institutional support

In India, many government and non-government organizations have had the focused attention on improving the medicinal plants sector. Opportunities for funding have been created to assist the person who is willing to work and to build capacity of the medicinal plants sector. According to the mandate of NMPB, the projects may be submitted for funding within two major schemes: viz., a promotional scheme and a commercial scheme. Major institutions involved in funding projects to the medicinal plants research in India are listed in Table 10.3. The major thrust areas within the promotional scheme are:

1. Survey and inventory of medicinal plants,
2. In-situ conservation and ex-situ cultivation of selected medicinal plants,
3. Production of quality planting material,
4. Diffusion of knowledge through education and communication,
5. Promotion of global and domestic market system, and
6. Strengthening research, development and man power.

Within the commercial scheme, the major thrust areas are:

1. Bulk production of medicinal plants and ensuring supply of quality planting material,
2. Expansion of selected medicinal plants farming areas,
3. Value addition in harvesting, processing and marketing of medicinal plants, and
4. Developing innovative marketing mechanism.

c. Stake holder's participation

Local communities, especially weaker and marginalized groups or ethnic minorities need to be involved in planning, designing, development and implementation of the research activities and learning studies in the project. The NGOs and GOs should consult and work with community- based organizations and engage them into participatory process to involve collectors, producers and traders including ultimate users, women and disadvantaged groups in project implementation. In each of the selected villages or communities, stakeholder represented CBO, NGO or PRI should implement and/or supervise the execution of projects. Their involvement from the very beginning of the project development process is expected to enhance people's participation in the project and provide benefit to a wide range of users.

TABLE 10.3   Major Institutions Involved in Funding Projects to the Medicinal Plants Research in India [17]

Institutions	Funding for major areas in medicinal plants research
All India Council for Technical Education, AICTE	Management technology
Council for Scientific & Industrial Research, CSIR	Ecology, taxonomy, biochemistry, survey, documentation, cultivation, genetics, agro-technology, conservation
Defense Research & Development Organization, DRDO	Agro-technology, survey, documentation, conservation
Department of Biotechnology, DBT	Agro-technology, molecular biology, biochemistry, rural bio-technology
Department of Science and Technology, DST	Taxonomy, ecology, pathology, survey, propagation, documentation, cultivation, conservation
G.B. Pant Institute of Himalayan Environment and Development, GBPIHED	Survey, documentation, cultivation, conservation
Herbal Research and Development Institute, HRDI	Survey, documentation, nursery development
Indian Council of Agricultural Research, ICAR	Breeding, pathology, molecular biology, training the growers
Indian Council of Medical Research, ICMR	Breeding, value addition
Ministry of Environment & Forest, MoEF	Survey, documentation, conservation, management, ecological impact assessment, cultivation
National Bank for Agriculture and Rural Development, NABARD	Cultivation, marketing
National Medicinal Plants Board, NMPB [15]	Survey, documentation, cultivation, marketing, conservation
University Grant Commission, UGC	Ecology, biochemistry, survey, documentation

10.7.2   MARKETING PERSPECTIVE

In order to understand the complex market and marketing related issues, market-related information, scooping of problems and opportunities, successful case studies with 'good practice' tag need to be surveyed and studied to develop a marketable product portfolio. The tool proposed is value-chain or Production-to-Consumption & Marketing (PC&M) model. The outcome of these studies can be useful to plan equitable commercialization, identify potential small and micro enterprises, and assess available raw material resources and product mix.

10.8   THE EFFORTS BY DIFFERENT AGENCIES

The efforts by different agencies, to promote the MAPs in this region, can be summarized in the following manner:
- Assessment of local conditions and requirements;
- Assessing the demand for medicinal and aromatic crops;
- Training and capacity building;
- Role of supporting individuals and institutions;
- Development of relevant literature in local language;
- Emphasis on group approach;
- Organizing the growers;
- Cultivation through the organized sector and linking the unorganized farmers;
- Pricing mechanism and buy back arrangement with partner industry.

10.9   AROMATIC PLANTS SUITABLE FOR CULTIVATION IN EASTERN REGION

- Geranium
- Patchouli
- CN-5
- Jama rosa
- Java citronella *(Cymbopogon winterianus)*
- Lemon grass *(Cymbopogon fleuosus)*
- Mentha *(Mentha arvensis)*
- Palma rosa *(Cymbopogon martinii)*
- Tulsi *(Ocimum basilicum)*
- Vetiver *(Vetiveria zizanioides)*

Tables 10.4 and 10.5 show the selected list of MAPs.

TABLE 10.4 Few Medicinal Plants Found Locally in Eastern India and Their Industrial Products

Name of Plant	Industrial Products
Agnisikha or Uluchandan *(Gloriosa superba)*	Colchicine
Bandar kekoa *(Mucuna prurita)*	L-Dopa
Birina	Vetiveryl acetate
Birina *(Vetiveria zizanioides)*	Vetiverol
Bishalyakarani *(Atropa belladonna)*	Atropine
Bishyalakarani	Hyoscyamine
Chahgoch	Theobromine
Chahgoch	Theophylline
Chahgoch *(Camellia sinensis)*	Caffeine
Dalchini, Tejapat, etc. (Cinnamomum spp.)	Eugenol
Dhatura *(Datura stramonium)*	Atropine
Geranium *(Pelargonium graveolens)*	Geraniol
Haldhi *(Curcuma longa)*	Cucurmin
Javacitronala	Dimethyloctane
Javacitronala or Ganbirina	Citronellol
Javacitronala or Ganbirina *(Cymbopogon winterianus)*	Citronellal
Mekuri Kendu *(Strychnos nux-vomica)*	Strychnine
Nayantora *(Catharanthus roseus)*, Sarpagandha	Ajmalicine
Omita *(Carica papaya)*	Papain
Pani tengeshi *(Marselia minuta)*	Marsilin
Podina	Linalyl acetate
Podina *(Mentha citrata)*	Linalool
Podina (Mentha spp.)	Menthol
Sadabahar or Periwinkle	Vinristine
Sadabahar or Periwinkle *(Chtharanthus roseus)*	Vinblastin
Sarpagandha	Rescinnamine
Sarpagandha	Reserpine
Sarpagandha *(Rauvlofia serpentina)*	Ajmaline

Source: *http://assamagribusiness.nic.in/agriclinics/Entrepreneurship%20dev.%20through%20medicinal.pdf.*

TABLE 10.5   Medicinal Plants Commercially Suitable for Cultivation in Eastern Region

Botanical name	Common name	Parts used	Nature of demand	Gestation period	Cultivation location	Demand as in 2004–05 (MT)
Andrographis paniculata	Kalmegh	Whole	Both, Dom/Exp	Short	Region wide	2197.3
Asparagus racemosus	Shatawari	Roots	Both, Dom/Exp	Short	Region wide	16658.5
Bacopa monnieri	Brahmi	Entire plant and leaves	Both, Dom/Exp	Short	Region wide	6621.8
Boehaavia diffusa	Punarnava	Stem	Both, Dom/Exp	Short	Region wide	3373.2
Centella asiatica	Brahmi	Whole	Both, Dom/Exp	Short	Region wide	134.5
Chlorophytum arundinaceum/boerivillianum	Safed musli	Bulbs/tuberous roots	Both, Dom/Exp	Short	Region wide	NA
Gloriosa superba	Shankhpushpi	Roots	Export	Short	Region wide	100.5
Glycyrrhiza glabra	Liquorice or Jethi madhu	Roots	Both, Dom/Exp	Short	Region wide	1359.8
Gymnema sylvestre	Madhunashini	Leaves, roots	Both, Dom/Exp	Short	Region wide	80.70t
Phyllanthus amarus/ P. niruri	Bhoomi amla	Roots	Both, Dom/Exp	Short	Region wide	2985.3
Piper longum	Pippali	Fruits	Both, Dom/Exp	Short	Region wide	6280.4
Rauwolfia serpentina	Sarpagandha	roots	Export	Short	Region wide	588.7
Tinospora cordifolia	Giloe	Roots, stem and leaves	Export	Short	Region wide	2932.6
Withania somnifera	Ashwagandha	Dried roots	Both, Dom/Exp	Short	Region wide	9127.5

Source: *https://idl-bnc.idrc.ca/dspace/bitstream/123456789/27286/9/g-120936.pdf*

10.10   BIO-PARTNERSHIP TO LINK RURAL COMMUNITIES WITH INDUSTRY

Links established with growers, traders, processors and consumers at different levels in a value-chain or production-to-consumption system framework. Equitable Bio-partnership arrangements between processing and marketing; health-care companies and community-based organizations developed to ensure dependable markets for the producers and quality supply for the industry. Industries have shown interest in direct collaboration with producer groups/associations and many are committed to a fair and ethical commercialization.

List of some important Indian pharma companies relevant to herbal products is shown below [15]:

- Albert David Ltd.
- Alembic Chemical Works Co. Ltd.
- Biological E. Ltd.
- Cadila Healthcare Pvt. Ltd.
- Cadila Pharmaceuticals Ltd.
- Cipla Ltd.
- Concept Pharmaceuticals Ltd.
- Croslands
- Dabur pharmaceutical Ltd.
- Deepharma Ltd.
- Dey's Medical Stores (Mfg.) ltd.
- Dr. Reddy's Lab.
- East India pharmaceuticals Ltd.
- Elders Pharma
- Franco-Indian Pharmaceuticals Ltd.
- Gufic Ltd.
- J. B. Chemicals Pharmaceuticals Ltd.
- Lupin Laboratories Ltd.
- Lyka Labs. Ltd.
- Merind Ltd.
- Parke-Davis (India) Ltd.
- Rallies India Ltd.
- Ranbaxy Laboratories Ltd.

- Themis (Phytomedica)
- TTK Pharma Pvt. Ltd.
- Wockhardt Ltd.

10.11 RESPONSE FROM THE FARMERS AND THE GOVERNMENT

Presently with support of ATMA, Patna and many other such agencies, more than about 2500 acres of land has been brought under the cultivation of MAPs in Bihar state alone. The number of farmers and area under these crops is likely to increase in future as a large number of farmers are attracted towards this sector in this region. Several civil society organizations have started working for promotion of this sector in this region, and the governments have also understood the need and importance of this sector and serious efforts are now being made to not only start their cultivation but also provide adequate financial support through commercial banks, provide quality planting material to the growers and provide market linkages, so that the growers are able to reap the benefits. Some Growers Associations are quite active in promoting cultivation of medicinal plants for enhancing livelihoods.

10.12 RECOMMENDATIONS FOR DEVELOPING THE MEDICINAL AND AROMATIC PLANTS SECTOR

The present worldwide interest in plant-based medicines of Indian origin needs to be harnessed by reframing a clear policy for the promotion of commercial cultivation, research and development, and for the increase in exports of medicinal plants. For the development of the medicinal plant sector, there is a need to develop the coordinated efforts at each stage (e.g., research, cultivation, collection, storage, processing, manufacturing and marketing), which would be supported by an appropriate policy framework. Some problems and their remedies for the medicinal plants-based economic venture identified are given in Table 10.6.

TABLE 10.6   Assessment of Problems and Remedies for Medicinal Plant Based Economic Venture in the Eastern India [17]

Activity	Problems	Possible remedy
Cultural system	Adoption of traditional medicinal knowledge on preparing herbal medical formulations is declining through generations.	Incentives should be given to the traditional herbal healers for preparation of herbal formulations, and attempts should be made to organize them.
	Traditional knowledge on many less known medicinal plant species has declined rapidly.	Documentation of such less known medicinal plant species should be made without any further delay.
Collection	Continued illegal collection from wild has led to depletion of many important species.	Enforcement of existing Acts (e.g., Wildlife Protection Act, Forest Act, Biodiversity Act, etc.).
	Mostly collected and processed by un- trained persons.	Training should be given for collection and processing.
	Competition for over-stocking has led to over-harvesting.	Large-scale farming of medicinal plants should be promoted.
Cultivation	Agro-technology is not available for many valuable medicinal plant species.	Development of agro-technology and promotion of rural bio-technology for large scale cultivation of economically important species also.
	Development of agro-technology is mainly focused on the low productive and high cost rare and endangered medicinal plant species.	Farmers should be encouraged by providing incentives, training and awareness on the latest developments and policies related to the medicinal plants.
		Selection of planting material for cultivation should be based on their habitats, locality, climate and elevations.
	High risk in farming, long gestation period, and low prices of medicinal plants discourage farmers to cultivate medicinal plants.	Introduction of mixed cropping system to reduce the risk
	Issuing license or permit to farmers for growing medicinal plants is a time consuming process, and farmers are sometime not aware of the process.	Process of issuing permits for cultivation of medicinal plants should be made easier and faster.

TABLE 10.6   (Continued)

Activity	Problems	Possible remedy
	Small and scattered land holdings of the farmers, and cultivation is restricted to small plots near the farmer's houses.	Restoration of barren lands and allocation of land at one place based on farmer's choice and consensus.
	Unavailability or low availability of irrigation facility	Rain water harvesting and construction of check dams on rivers and rivulets for irrigation purposes.
	Lacking of linkages among different stakeholders.	Development of capacity building programs for all stakeholders.
Role of Biotech-nology	Low success rate in developing planting materials.	Need of in-depth research to enhance the rate of success.
	Low yield unable to meet the commercial needs.	Development of high yielding varieties.
Marketing	The supply chain of medicinal plants is quite large and primary producers are dependent on the middlemen and still they face difficulty in selling the product.	Direct selling to industry by producers should be encouraged. Buy-back arrangements between farmers and pharmaceutical companies might be useful.
	Improper sharing of benefits due to lack of awareness among farmers and herb collectors on the real prices of medicinal plant.	Need of diffusion of information by distribution of pamphlets and conducting awareness programs on various aspects of medicinal plants.
	Lacking of well-planned marketing infrastructure for medicinal plants.	Development of infrastructure with the help of various stakeholders including medicinal plants board.
Bio-prospecting	Low awareness on the values of resources and traditional knowledge.	Documentation of traditional knowledge on medicinal plants and their uses.
	The younger generations of herbal practitioners are not keen to adopt the tradition as a profession.	Renew the available herbal formulations by standardizing their efficacy, and to establish a Social Capital Trust for herbal practitioners in order to promote the tradition.

TABLE 10.6   (Continued)

Activity	Problems	Possible remedy
	Unequal distribution of profits to the low profile stakeholders such as farmers and herb gatherers.	Sharing of benefits should be on the basis of labor and efforts.
Conserva-tion	Essential health commodity and maximum dependency on wild stock.	Setting up medicinal plants con-servation areas.
	Encroachment by outsiders and ille-gal collection from wild.	Enforcement of Forest and Wild-life Protection (Acts).

10.13   CONCLUSIONS

Selection of medicinal plant species for cultivation is an initial impor-tant step for the development of the medicinal plants sector. Economic feasibility is the major rationale for a decision to bring medicinal plant species into cultivation. Apart from the priority species selected by the Planning Commission [17] and the NMPB, the rare species banned for collection from the wild should also be taken on a priority basis for culti-vation because a majority of such species are very expensive, have high demand and low supply.

Cultivation may not be economical if a medicinal plant species is abun-dant in the wild and easily collected. Therefore, the less abundant species in the wild should be promoted for the large-scale cultivation. Farming of any medicinal plant species should be brought into practice only after the reliable cultivation technology of the concerned species is available. A large variation in climatic and soil conditions in eastern India sustain a variety of medicinal plant species, which may be cultivated according to their niche. For developing the medicinal plants sector, there is an urgent need to:

a.   Document indigenous uses of medicinal plants,
b.   Certify raw material for quality control,
c.   Develop and improve the agro-technology for valuable medicinal plants,
d.   Officially recognize and protect the customary laws of indigenous people,
e.   Prepare a clear policy for granting permits for cultivation within stipulated time,

f. Conduct regular research and training on better harvesting and processing techniques,

g. Investigate various pathological agents infecting medicinal plants,

h. Setup a community-based management of medicinal plants farming and marketing,

i. Analyze the market policies,

j. Monitor and evaluate the status of medicinal plants with the assistance of local communities,

k. Conserve the critical habitats of rare medicinal plant species, and

l. Share benefits judiciously arising from local people's knowledge on medicinal plants.

These attempts may reduce dependency on wild resource base, and generate alternative income opportunities for the rural and underprivileged communities.

The medicinal plants sector can be improved if the agricultural support agencies would come forward to help strengthen the medicinal plants growers and if research institutions would help the plant growers by improving their basic knowledge about cultivation practices. Awareness and interest of farmers, supportive government policies, assured markets, profitable price levels, access to simple and appropriate agro-techniques, and availability of trained manpower are some of the key factors for successful medicinal plants cultivation.

The diffusion of any available scientific knowledge on medicinal plants should be made operational by a network structure of communication. Currently there are number of herbs which are used in curing diseases but are not documented in details due to a lack of communication and relatively low frequency of their uses. The traditional uses of low profile and lesser-known medicinal plants should also be documented to disseminate their therapeutic efficacy by preparing well acceptable medicines and also to reduce the pressure on over-exploited species.

On many occasions, the collection of planting material, especially of rare and endangered medicinal plant species from natural habitats for various experimental purposes by researchers, also poses a threat on their natural populations in wild. The researchers must be aware on the germination potential, seedlings and rhizomes survival strategies of the desired species collected from wild for scientific experiments. Researchers must plant a similar number of individuals back in nature after completion of research work on the collected species. There is also a communication problem between researchers and farmers.

This communication problem limits a researcher's capability to deal with the farmers' problems. Hence, communication links between researchers, extension services of institutions, and farmers should be strengthened.

KEYWORDS

- AICTE
- CSIR
- DBT
- DRDO
- DSE
- DST
- Farmers
- GBPIHED
- HRDI
- ICAR
- ICIMOD
- ICMR
- MADPs
- MAPPA
- MAPs
- medicinal plants
- MoEF
- NABARD
- natural habitat
- NMPB
- PC&M
- Planning Commission
- planting material

- rhizomes
- seedlings
- WHO
- WOCMAP
- WTO

REFERENCES

1. Bordeker, G. (2002). Medicinal plants: Towards sustainability and security. A paper prepared for IDRC Medicinal Plants Global Network, IDRC, South Asia Regional Office, New Delhi, India.
2. CRPA, (2001). Demand Study for Selected Medicinal Plants: Volumes I & II. Center for Research, Planning and Action (CRPA) for Ministry of Health & Family Welfare, GOI, Department of ISM&H & WHO, New Delhi, India.
3. Farooquee, N. A., Majila, B. S., & Kala, C. P. (2004). Indigenous knowledge systems and sustainable management of natural resources in a high altitude society in Kumaun Himalaya, India. Journal of Human Ecology, 16, 33–42.
4. ICAR Research Complex for Eastern Region, Patna, various Annual Reports (2001–2008).
5. ICAR Research Complex for Eastern Region, Patna, Status Paper 2001.
6. Kala, C. P. (2004). Studies on the Indigenous Knowledge, Practices and Traditional Uses of Forest Products by Human Societies in Uttaranchal State of India. Almora: GB Pant Institute of Himalayan Environment and Development.
7. Kala, C. P. (2004). Revitalizing traditional herbal therapy by exploring medicinal plants: A case study of Uttaranchal State in India. In: Indigenous Knowledge's: Transforming the Academy. Proceedings of an International Conference. Pennsylvania: Pennsylvania State University. pp. 15–21.
8. Kala, C. P., Pitamber Prasad Dhyani, & Bikram Singh Sajwan (2006). Developing the medicinal plants sector in northern India: challenges and opportunities. Accessed on 13/02/2009, Journal of Ethnobiology and Ethnomedicine, 2, 32. http://www.pubmedcentral.nih.gov/articlerender.fcgi?artid=1562365.
9. Kanel, K. R. (2000). Sustainable Management of Medicinal and Aromatic Plants in Nepal: A Strategy. A Study Commissioned by IDRC/SARO, Medicinal and Aromatic Plants Program in Asia (MAPPA). New Delhi, India.
10. Karki, M. B., & Nagpal, A. (2004). Marketing Opportunities and challenges for Medicinal Aromatic and Dye plants (MADPs). A position paper presented at the International Workshop on Medicinal Herbs and Herbal Products' Livelihoods and Trade Options How to Make Market Work for Poor?

11. Karki, M. (2001). Certification and marketing strategies for sustainable commercialization of medicinal and aromatic plants in Chhattisgarh. Paper presented at the National Research Seminar on Herbal Conservation, Cultivation, Marketing and Utilization with special emphasis on Chhattisgarh, The Herbal State, December 13–14, Raipur, India.

12. Karki, M. B., Brajesh Tiwari, Badoni, A., & Bhattarai, N. (2003). Creating livelihoods enhancing medicinal and aromatic plants based biodiversity – rich production systems: Preliminary lessons from South Asia. Paper Presented at The 3rd World Congress on Medicinal and Aromatic Plants for Human Welfare (WOCMAP III). 3–7 February, Chiang Mai, Thailand.

13. Karki, M. B. (2002). Organic conversion and certification: a strategy for improved value-addition and marketing of medicinal plants products in the Himalayas. Paper presented at the Regional Workshop at Wise Practices and Experimental Learning in the Conservation and Management of Himalayan Medicinal Plant; December 15–20, Kathmandu, Nepal.

14. Karki, M. B. (2000). Commercialization of natural resources for sustainable livelihoods: the case of forest products. Paper presented at the International Conference on Growth, Poverty Alleviation and Sustainable Management in the Mountain Areas of South Asia organized by German Foundation for International Development (DSE) and International Center for Integrated Mountain Development (ICIMOD); January 31–February 4, Kathmandu, Nepal.

15. National Medicinal Plants Board Publication (2002). Cultivation Practices of Some Commercially Important Medicinal Plants.

16. Nautiyal, S., Rao, K. S., Maikhuri, R. K., Negi, K. S., & Kala, C. P. (2002). Status of medicinal plants on way to Vashuki Tal in Mandakini Valley, Garhwal, Uttaranchal. Journal of Non-Timber Forest Products, 9, 124–131.

17. Planning Commission (2000). Report of the Taskforce on Medicinal Plants in India. Planning Commission, Government of India, Yojana Bhawan, New Delhi, India, 16–17.

18. Rawat, R. B. S., & Uniyal, R. C. (2004). Status of medicinal and aromatic plants sector in Uttaranchal: initiatives taken by the Government of India. Financing Agriculture, 36, 7–13.

19. Regmi, S., & Bista, S. (2002). Best practices in collection and cultivation of medicinal plants for sustainable livelihoods in Himalayan communities. Paper presented at the Regional Workshop at Wise Practices and Experimental Learning in the Conservation and Management of Himalayan Medicinal Plant; December 15–20, Kathmandu. CECI, Nepal.

20. Singh, K. M., & Jha, A. K. (2008). Medicinal and Aromatic Plants Cultivation in Bihar, India: Economic Potential and Condition for Adoption. Available at http://dx.doi.org/10.2139/ssrn.2213308.

21. Tiwari, B. K. (2002). Traditional system of Bay Leaf (C. Tamala) management by War Khasi community of Meghalaya, India. Paper presented at the Regional Workshop at Wise Practices and Experimental Learning in the Conservation and Management of Himalayan Medicinal Plant; December 15–20, Kathmandu, Nepal. Center for Environmental Studies, North-Eastern Hill University, Shillong, India.

CHAPTER 11

GREEN SYNTHESIS OF AGNPS FROM *SPINACIA OLERACEA* LEAVES AND THEIR ANTI-MICROBIAL EFFICACY AGAINST HUMAN PATHOGENIC MICROBES

MANBIR KAUR, BRAHMDUTT ARYA, and VEDPRIYA ARYA

CONTENTS

11.1 INTRODUCTION

Nanotechnology is the most fascinating area of research in the field of material science. Currently, it involves various physical and chemical methods for nanoparticles synthesis. But the main problem with these methods is the production of toxic byproducts, shows that these are not environmentally safe methods. Thus there is a growing need for 'green chemistry' that includes a clean, nontoxic and environmental friendly methods of nanoparticles synthesis. For environmental safety concerns, researchers in the field of nanoparticles synthesis and assembly have turned to the biological systems [1].

Biological synthesis of nanoparticles proved to be cost effective over chemical means as it does not involve physical barriers with regard to reducing agents and eliminates the toxic effects of chemicals used for the synthesis. At present, a number of living organisms are already known to synthesize nanoparticles such as cyanobacteria, bacteria, fungi, acitomycetes, biomolecular and various plant materials such as *Cinnamomum camphora, Medicago sativa, Pelargonium graveolens, Avene sativa, Azardirchta indica, Tamarindus indica, Parthenium hysterophorus, Tritium vulgare, Acanthella elongata, Sesuvivm potulacastrum* and gold nanoparticles also synthesized by biomolecules like Honey. Leaf extracts of Neem, Geranium, Hibiscus, Cinnamon, Tamarind, Coriander, Brassicaceae spp. and many plant and seeds such as Gram, Maize, Moong have been used for development of nanoparticles. So, the living plants are considered as eco-friendly nanofactories [1, 7].

In the present study, *Spinacia oleracea* leaf extract is used as reducing agent of silver nitrate into silver nanoparticles. This plant is easily available and edible in all over the world. Its leaves are chiefly utilized for making various dishes in several cuisines. Juice of leaves is recommended for various disorders. This plant is used to produce silver nanoparticles and these nanoparticles were subjected to anti-microbial action study against several pathogens.

11.2 METHODOLOGY

11.2.1 PREPARATION OF AQUEOUS LEAF EXTRACT

The plant matter is collected from the local vegetable market and specimen is identified with the herbarium of G.N.G. College (GNG 1103). The leaves are thoroughly washed and oven dried at 50°C. The dried leaves are crushed and an aqueous extract is prepared by adding 10 times of water by both hot and cold percolation methods. The extra water is evaporated on a water bath and a dark, viscous solution (Figure 12.1).

FIGURE 11.1 Plant sample, extract and color changes Spinacia oleracea leaf extract after AgNO$_3$ treatment at different temperatures.

11.2.2 BIOSYNTHESIS OF SILVER NANOPARTICLES

For each experimental set up, two flasks were prepared, in the first flask, 2.5 mL of extract with 50 mL of 1 mM AgNO$_3$ was added and no leaf extract was added in the second flask and considered as control. Separate experiments were done for different reaction conditions like pH variation [9–11], temperature variation (–20°, 0°, 37°, 100°C) and extract variation (0.5, 2.5, 4.5 mL).

11.2.3 CHARACTERIZATION OF SILVER NANOPARTICLES

a. **UV-Vis Analysis (Preliminary Analysis)**
   UV-Vis spectra were recorded at 450 nm and absorbance was plotted on graph by Graph pad prism 5.0 software.

b. **Transmission Electron Microscopy (TEM) (Confirmatory Analysis)**

Confirmatory analysis was carried out by TEM analysis. Pellets were prepared by centrifugation suspension at 10,000 rpm for 20 minutes. The pellets were carried to Sophisticated Analytical Instrumentation Facility (SAIF), Panjab University, Chandigarh and TEM analysis was carried out.

c. **Evaluation of Anti-Bacterial Activity**

The pathogenic bacterial strains were isolated from the diseased patients of Adesh Institute of Medical Science and Research, Bathinda. The anti-microbial activities of the silver nanoparticles were determined by agar well diffusion method according to NCCLS standards [6].

11.3   RESULTS AND DISCUSSION

The samples when treated with different reaction conditions, change in color of extracts suspension from dark green to brownish red were observed. This preliminary color change showed the presence of silver nanoparticles or reduction of Ag^+ of $AgNO_3$ to Ag°. After observing changes in color of the extracts, they were scanned from 420–680 nm in and maximum absorbance was observed at 450 nm due to Surface Plasmon Resonance (SPR) of silver nanoparticles.

11.3.1   UV-VIS STUDIES

11.3.1.1   Effect of Temperature

It was concluded that plant extracts incubated at 100°C temperature from hot percolation treatment have more number of silver nanoparticles as compared to those from cold percolation treatment results. It was observed with increase in temperature from –20°C to 100°C, increase in the number of silver nanoparticles was observed due to their property of SPR. In accordance to above reported results, Dubey and his coworkers in 2010 [4] used temperature variations from 25°C to 150°C and noted increase in peak sharpness due to increasing temperature and concluded that increase in absorbance peak sharpness may be due to size of the synthesized nanoparticles with high temperature, particle size may be small which results into sharpness of plasmon resonance band of Silver nanoparticles. Similarly, Song and his coworkers [10] reported that maximum numbers of silver nanoparticles were synthesized after 10 hours of incubation.

11.3.1.2   Effect of pH

It was interpreted that extracts with pH of 11 prepared from hot percolation method produced more number of silver nanoparticles as compared to those from cold percolation method. It was observed that with increase in pH conditions there was change in color and absorbance increased due to SPR. In accordance to above reported results, Dubey and his coworkers [4] and many others had shown that alkaline pH solutions proves stability in synthesis of nanoparticles as compared with acidic pH solutions. Also it can be said that at acidic pH, the particle size are comparatively larger than the basic pH, as blue shift clearly reported in the SPR spectra. In contrast to the above results, Castro and his coworkers [2] reported that on pH 11, less number of nanoparticles was produced than on pH 10.

11.3.1.3   Effect of Concentration Variation

It was interpreted that extracts with raw extract concentration 4.5 mL prepared from hot percolation method produced more number of silver nanoparticles as compared to those from cold percolation method. It was observed that with increase in raw plant extract concentration there was change in color and absorbance increased due to SPR. Sathish Kumar and his coworkers [8] showed that there was increase in absorbance with respect to time and the dosage of *Cinnamon zeylanicum* bark powder (CBP) and CBPE (extract), respectively, was noticed in UV-Vis spectrum.

11.3.2   *TRANSMISSION ELECTRON MICROSCOPY (TEM) ANALYSIS*

TEM analysis is confirmatory technique applied for characterization of silver nanoparticles synthesized from plant extracts. Before performing TEM analysis, samples from variants with maximum absorbance were selected that is extracts from temperature, pH, time, extract concentration showing maximum absorbance were selected for TEM analysis. From these extracts pellets were prepared and using drop method TEM images were obtained. TEM images showed that nanoparticles produced are mostly spherical in shape and their size varies from 10 to 12 nm. The Magnified images were observed by transmission electron microscope (TEM) as shown in Figure 11.2.

11.3.3   *ANTI-BACTERIAL EFFECT OF SILVER NANOPARTICLES*

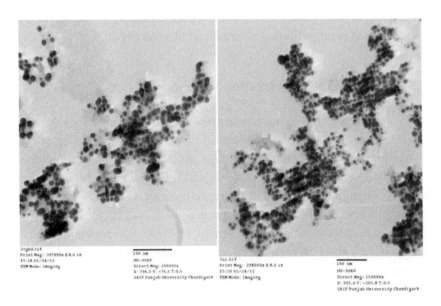

FIGURE 11.2   TEM Micrographs showing Ag nanoparticles.

The anti-bacterial activity of silver nanoparticles isolated from spinach leaf extract was tested against four clinically isolated pathogens *Enterococcus, Pseudomonas aeruginosa, E. coli* and *Klebsella pneumonia* by using agar disk diffusion assay method. These isolated silver nanoparticles showed significant anti-bacterial activity against *Enterococcus* and *Pseudomonas aeruginosa* (14 mm zone of inhibition each) as compared to *E. coli* and *Klebsella pneumonia.*

11.4 CONCLUSIONS

Silver nanoparticles (AgNPs) have promising potential in agriculture, biomedicine, optics, environment and health care sectors. A novel, eco-friendly mechanism of green synthesis of silver nanoparticles from the aqueous extract of *Spinacia oleracea* leaves was adopted. UV-visible spectrophotometric study showed the characteristic peak for AgNPs at wavelength 430 nm. The optical density at 430 nm increased after addition of plant leaf extract, indicating increase in formation of nanoparticles. Comparative time course and pH variation analysis for AgNP synthesis carried out at different reaction temperatures ($-20°$, $4°$, $37°$ and $100°C$) and pH (9, 10, 11) revealed the maximum synthesis of nanoparticles at $100°C$ and pH 11. Electron microscopy analyzes confirmed the presence of well dispersed AgNPs, predominantly with spherical shapes and in 10–12 nm size range. The biologically synthesized AgNPs showed significant antimicrobial effect against clinically isolated strains of *Enterococcus fecalis, Pseudomonas aeruginosa, E. coli* and *Klebsella pneumonia.* These observations lead to an ecofriendly approach of AgNPs synthesis for the management of clinical pathogens. These nanoparticles are environmental friendly and completely safe. Moreover, this biosynthesis approach is close to principles of 'Nature.' Biological synthesis of nanoparticles involves natural phenomenon that take place in the biological systems. At present, there are several evidences of nanoparticles formation in Algae, Fungi, Yeast and higher plants. In the present study leaf extract of papaya was used to synthesize silver nanoparticles and their antibacterial effect was tested against human pathogenic bacteria. Papaya plant has a great medicinal value. So, these research findings enlighten the future potentialities of taking a new look at old systems that retain some of the most fascinating Nature's secrets. Moreover these studies provide the possibility of an environmental friendly method to remediate mining wastes.

KEYWORDS

- AgNPs synthesis
- algae
- biological synthesis
- *E. coli*
- fungi
- green synthesis
- higher plants
- nanoparticles
- nanotechnology
- papaya
- silver nanoparticles
- *Spinacia oleracea*

- temperature variation
- yeast

REFERENCES

1. Arya, V. (2010). Living Systems: Eco-Friendly Nanofactories. Digest Journal of Nanomaterials and Biostructures, 5(1), 9–21.
2. Castro, L., Blázquez, M. L., Munoz, J. A., González, F., Balboa, C. G., & Ballester, A. (2011). Process Biochemistry, 46, 1076–1082.
3. Chandran, S. P., Chaudhary, M., Parricha, R., Ahmad, A., & Sastry, M. Biotechnol. Progress, 22, 577.

4. Dubey, S. P., Lahtinenb, M., & Sillanpaa, M. (2010). Process Biochemistry, 45, 1065–1071.
5. Gardea-Torresdey, J. L., Gomez, E., Peralta-Videa, J. R., Parsons, J. G., Troiani, H., & Jose-Yacaman, M. (2003). Langmuir, 19, 1357–1361.
6. Kiehlbauch, J. A., Hannett, G. E., Salfinger, M., Archinal, W., Monserrat, C., & Carlyn, C. Use of national committee for clinical laboratory standards guidelines for disk diffusion susceptibility testing in New York state laboratories, J. Clin. Microbiol., 38, 3341–3348.
7. Komal, R., & Arya, V. (2013). Biosynthesis and characterization of silver Nanoparticles from aqueous leaf extracts of Carica papaya and its antibacterial activity. International Journal of Nanomaterials and Biostructures, 3(1), 17–20.
8. Sathishkumar, M., Sneha, K., Won, S. W., Cho, C. W., Kim, S., & Yun, Y. S. (2009). Colloids surfaces B. Biointerfaces, 7, 332–338.
9. Singaravelu, G., Arockiamary, J. S., Ganesh Kumar, V., & Govindaraju, K. (2007). Colloids and interfaces B. Biointerfaces, 57, 97–101.
10. Song, J. Y., Jang, H. K., & Kim, B. S. (2009). Process Biochemistry, 44, 1133–1138.

PART 5

RENEWABLE ENERGY USE IN AGRICULTURE

CHAPTER 12

A COMPARATIVE STUDY OF BIOMASS MATERIALS USING DOWNDRAFT GASIFIER

P. K. PRANAV, THANESWER PATEL, and Y. K. RAO

CONTENTS

12.1   INTRODUCTION

Increasing demand of energy in the developing countries emphasizes the scientist and researcher to find acceptable alternative sources of energy, independent from the petroleum product. Renewable energy is the well-proven source of energy which is widely available throughout the universe. Out of various source of renewable energy, biomass is a well kwon source of energy which is being used by human from the existence of mankind. Biomass can be defined as *"all the matter that can be derived directly or indirectly from plant photosynthesis."*

Biomass gasification is a thermo-chemical process of conversion of biomass into the mixture of combustible and non-combustible gases by partial oxidation at high temperature (800–900°C) in the presence of a gasifying medium such as air, oxygen or steam. The gas produced from biomass in known as synthesis gas (syngas), which is a mixture of carbon monoxide (CO), carbon dioxide (CO_2), hydrogen (H_2), water (H_2O) and a small amount of methane (CH_4).

A downdraft gasifier is well-established device for biomass gasification, however, performance of gasifier and their economic viability dependents on type of biomass, i.e., feedstock. The use of syngas for power generation is widely accepted and considered mature technology.

There is a huge scope of syngas generation from biomass in Arunachal Pradesh as 82% of total geographical area (83743 km²) is under forest cover. The economy of gasification from mainly depends on availability of biomass, non-commercial use and low cost. The state is very rich in such kind of biomass which is either being used as firewood or of no use. Numerous works on bamboo gasification [1–6] and wood gasification are reported [7–20], but none has tried for the local available feedstock in Arunachal Pradesh. Keeping these arguments in view, a research study was formulated to compare the gasification potential of three locally available feedstocks using downdraft gasifier.

12.2   MATERIALS AND METHODS

12.2.1   BIOMASS FEEDSTOCK

The selection was carried out on the basis of availability in local area, non-commercial use, low cost and frequent pruning. On these basis three feedstocks were selected namely Khokan (*Duabanga grandiflora*), Kadam (*Anthocephalus chinensis*) and Bamboo (*Poaceae*).

12.2.2   MEASUREMENT OF PHYSICAL PROPERTIES

The selected feedstocks were cut into small pieces (called chips) using wood cutter as shown in Figure 12.1. The various physical properties such

Bamboo chips   Khokan wood chips   Kadam wood chips

FIGURE 12.1   Selected feedstock chips used for comparative study.

as size, shape, moisture content, bulk density, ash content, volatile matter content, fixed carbon percentage, calorific value and heating value of the feedstocks were measured/calculated which is presented in Table 12.1. Ash content was determined in muffle furnace as per the ASTM D1102–84 standards. Further, volatile matter content was also measured as per ASTM E1755–01. The Fixed carbon was calculated by the following equation:

$$\text{Fixed carbon (\%)} = 100 - [(\% \text{ volatile matter} + \% \text{ Ash})] \qquad (1)$$

The calorific value of a fuel is the quantity of heat produced by its combustion at constant pressure and under normal conditions. It was determined with the help of bomb calorimeter.

TABLE 12.1   Physical Properties of Selected Feedstock

Parameters	Feedstock		
	Bamboo	**Khokan**	**Kadam**
Ash content (%)	4.0	3.15	2.65
Bulk density (kg/m³)	305	360	330
Calorific value (kcal/kg)	3831	3900	3568
Fixed carbon (%)	8.75	14.60	12.35
Moisture content (% wb)	9.95	13.69	14.24
Volatile matter (%)	87.25	87.75	85.00
Calculated higher heating value (MJ/kg)	16.66	18.04	19.60

12.2.3   EXPERIMENTAL PROCEDURE

The gasification of selected feedstocks was carried out in a commercial fixed bed downdraft biomass gasifier plant (Model: WBG-10) procured from M/s Ankur Scientific Energy Technologies Pvt. Ltd., Baroda (Gujarat). The gasification was performed at 4 levels of air flow for all the three feedstocks. All the experiment was replicated thrice. The performance parameters measured during the experiment are feed rate, air flow rate, pressure at various places, flame and gas temperature and ash content. Further, produced gas was collected in balloons (Figure 12.2) for analyzing the gas composition in laboratory.

12.2.4   GAS COMPOSITION

The analysis of collected syngas was carried out at the *Centre for Energy at Indian Institute of Technology*, Guwahati, India using a Gas Chromatograph (GC). The balloon containing the producer gas mixture was connected to the inlet column and after observing the air bubbles in the outlet column which is placed in the water container, the start button was pressed. A graph plot of time verses voltage appears on the computer screen. The sample representative graph is shown in Figure 12.3.

FIGURE 12.2. Producer gas samples collecting in balloon.

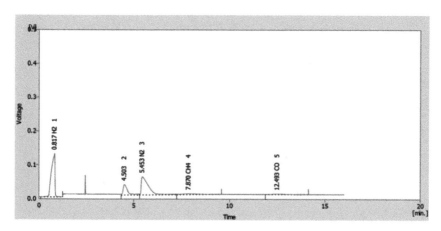

FIGURE 12.3. Sample graph coming out from GC.

The several peaks were observed on the graph, which represents the amount of gas components present in the producer gas. For example, first peak represent the H_2, second peak for N_2 etc. The actual percentage of gas was calculated by applying the different calibration factors for different gas compositions. The total time for each gas analysis was approximately 20 minutes.

12.3   RESULTS AND DISCUSSION

12.3.1   EFFECT OF AIR FLOW RATE ON FEED RATE

It was observed that the feed rate increases with increase in the air flow rate for all the feedstocks (Figure 12.4). It was observed that the feed rate was increased by 185.7, 183.1, and 183.3% when flow rate was increased from 2.5 to 6.5, 2.3 to 6.7 and 1.7 to 5.8 m³/h for bamboo, khokan and kadam feedstocks, respectively. The increase in the air flow rate provides higher oxygen to burn more amount of feedstock in the same time period.

12.3.2   EQUIVALENCE RATIO

Equivalence ratio reflects the combine effect of air flow rate, rate of feedstock supply and duration of run. The Equivalence ratio with respect to air

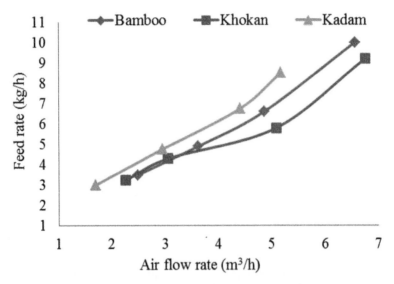

FIGURE 12.4 Effect of air flow rate on feed rate consumption.

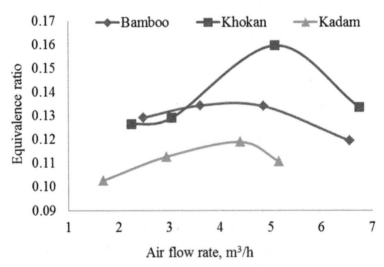

FIGURE 12.5 Effect of equivalence ratio with respect to air flow rate.

flow rate for all the feedstocks is given in Figure 12.5. It was found that the average equivalence ratio was 0.13, 0.14 and 0.11 for bamboo, khokan and kadam, respectively. As equivalence ratio characterizes the overall performance of gasifier [8], it was found that gasifier was performing better with khokan feedstock compared to bamboo and kadam feedstocks.

12.3.3 EFFECT OF AIR FLOW RATE ON PRODUCER GAS TEMPERATURE

The effect of air flow rate on producer gas temperature is shown in the Figure 12.6. It evident from the figure that gas temperature decreases as flow rate increases. The gas get more time to stay in reactor because of low flow rate may be a possible reason of higher temperature. Again, due to high gas temperature (>400°C) at low air flow, there is a partial combustion of producer gas which further increases the gas temperature. The gas temperature varies from 256–390, 225–450 and 252–355°C when air flow rate was varied from 3.05–6.58, 2.26–6.75 and 1.69–5.17 m^3/h for bamboo, khokan and kadam, respectively. Similar results have also been reported by Rajvanshi [21] about the gas temperature.

12.3.4 EFFECT OF AIR FLOW RATE ON FLAME TEMPERATURE

The effect of air flow rate on producer gas flame temperature is shown in Figure 12.7. It was witnessed that the flame temperature increased with air flow rate and gets maximum value between ½ to 3/4th opening of gas flow rate control valve for all the feedstocks. With further increase in flow rate, flame temperature was found to decrease significantly because, atmospheric air gets less time to convert into producer gas due to high flow rate.

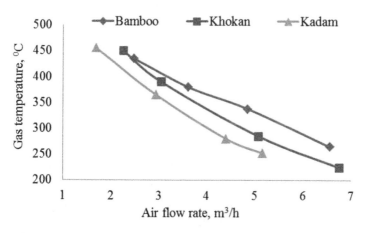

FIGURE 12.6   Effect of air flow rate on producer gas temperature.

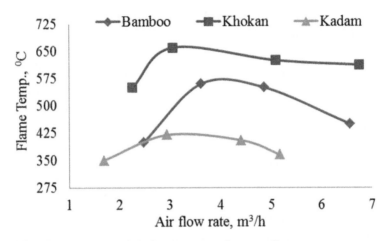

FIGURE 12.7   Effect of air flow rate on producer gas flame temperature.

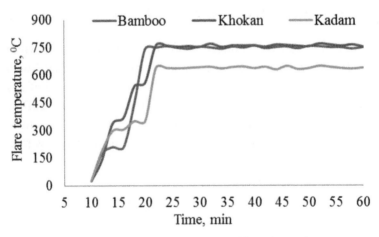

FIGURE 12.8   Variation of flame temperature with respect to time.

The maximum flame temperature was 560, 660 and 420°C in bamboo, khokan and kadam, respectively.

A typical curve showing the variation of flame temperature with respect to time at 3/4ᵗʰ opening of the control valve for various feedstocks is given in Figure 12.8. It was observed that in case of bamboo and khokan flame temperature reached up to 750°C however in case of kadam it reached up to 650°C.

12.3.5   EFFECT OF AIR FLOW RATE ON GAS COMPOSITION

The percentage of syngas composition in case of all feedstocks at various flow rates is discussed separately for hydrogen (H_2), carbon-monoxide (CO), carbon dioxide (CO_2), and nitrogen (N_2).

12.3.5.1   Hydrogen and Carbon Monoxide

It was observed that percentage of H_2 and CO present in the gas increased with increase in the air flow rate and gets a maximum value between ½ to 3/4th opening of gas flow control valve for all the feedstocks (Figures 12.9 and 12.10). Increasing the gas flow rate improve the bed temperature profile leading to higher conversion of non-combustible component (CO_2, H_2O) in to combustible component (CO + H_2).

Further, increase in gas flow rate resulted in decreased gas component because non-combustible does not get sufficient time for full conversion into combustible. The producer gas (CO + H_2) production was found 0.30, 0.32 and 0.23 Nm^3/kg for bamboo, Khokan and kadam, respectively.

12.3.5.2   Carbon Dioxide

It was observed that composition of CO_2 percentage varies from 10–12, 11.5–12 and 10–12% for khokan wood chips followed by kadam and

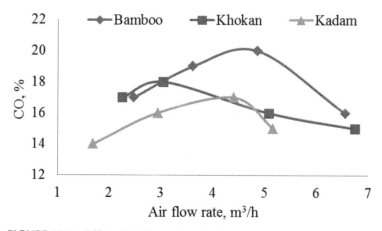

FIGURE 12.9   Effect of air flow rate on hydrogen content.

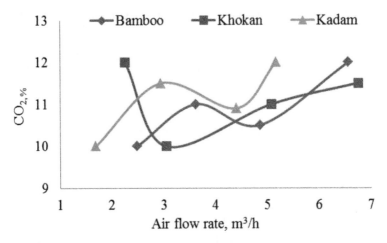

FIGURE 12.10 Effect of air flow rate on carbon monoxide content.

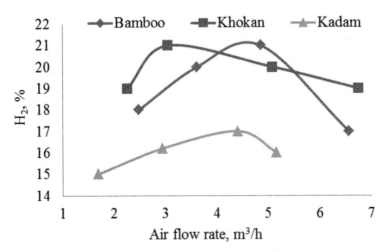

FIGURE 12.11 Effect of air flow rate on carbon dioxide content.

bamboo wood chips, respectively. Further, no significant difference in CO_2 was observed with respect to air flow rate for all the feedstocks (Figure 12.11).

12.3.5.3 Nitrogen

The effect of air flow rate on nitrogen composition is shown in Figure 12.12. It was observed from the graph that the nitrogen percentage increased

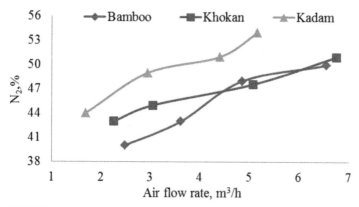

FIGURE 12.12   Effect of air flow rate on carbon dioxide content.

with increased in the air flow rate for all the feedstocks. It was found that the N_2 increased by 27.5, 18.6 and 22.7% when flow rate increased from 3.05–6.58, 2.26–6.75 and 1.69–5.17 m³/h for bamboo, Khokan and Kadam, respectively.

12.4   SUMMARY

Northeast region of India is known for its diverse and the most extensive lush forest coverage and rich in natural resources, including various plants based biomasses. Biomass gasification is a promising technology for conversion of solid biomass to gaseous fuel, which displaces the use of fossil fuels and reduces CO_2 emissions. Therefore, a comparative study of gasification for three locally available feedstocks namely khokan, bamboo and kadam, was carried out using down draft gasifier. Gas chromatograph was used for gas composition analysis viz., H_2, CO, CO_2 and N_2. The physical and chemical characteristics showed that khokan has higher calorific value (3900 kcal/kg) compared to bamboo (3881 kcal/kg) and kadam (3568 kcal/kg). The feed rate was found to increase by 185.7, 183.1 and 183.3% with the increase in flow rate from 2.5 to 6.5, 2.3 to 6.7 and 1.7 to 5.8 m³/h for bamboo, khokan and kadam feedstocks. The average equivalence ratio, which reflects the combined effects of air flow rate, feed rate and duration of run, was 0.13, 0.14 and 0.11 for bamboo, khokan and kadam, respectively.

The producer gas composition ($CO + H_2$) was found the maximum at ½ to 3/4th opening of the air flow control valve for all feedstocks. Further, higher value of same was found for khokan ($CO = 18\%$ and $H_2 = 21 \%$) followed by bamboo ($CO = 20\%$ and $H_2 = 21 \%$) and kadam ($CO = 17\%$ and $H_2 = 17\%$). Based on these results, it was concluded that khokan feedstock performed well in the gasifier at air flow rate of 5 m³/h compared to others selected feedstock. The feed rate and gas production rate were 5.8 kg/h and 0.32 Nm³/kg, respectively for the khokan feedstock.

KEYWORDS

- **air flow rate**
- **biomass feedstock**
- **carbon monoxide**
- **equivalence ratio**
- **gas composition**
- **gasifier**
- **hydrogen**
- **nitrogen**

REFERENCES

1. Scurlock, J. M. O., Dayton, D. C., & Hames, B. (2000). Bamboo: an overlooked biomass resource? ORNL/TM-1999/264. Oak Ridge National Laboratory, Oak Ridge, Tennessee. 34.
2. Ganesh, A. (2003). Bamboo characterization for thermochemical conversion and feasibility study of bamboo based gasification and charcoal making. IIT Mumbai. National Renewable Energy Laboratory, Subcontractor Report (NREL/SR-510-29952), Golden, CO, USA.
3. Sinha, P., & Bajpai, R. (2009). Plant it – bring good luck to people, energy, and environment. Akshay Urja (Renewable Energy), 2, 18–21.
4. Sinha, A. K., Barkakat, M., Nath, D., & Sharma, S. K. (2010). Economic viability of bamboo dust based gasification plant for a paper mill. International Conference on Renewable Energies and power Quality, Las Palmas de Gran Canaria (Spain), 13th to 15th April.

5. Chen, W. H., Du, S. W., Tsai, C. H., & Wang, Z. Y. (2012). Torrefied biomasses in a drop tube furnace to evaluate their utility in blast furnaces. Bioresource Technology, 111, 433–438.

6. Chiang, K. Y., Chen, Y. S., Tsai, W. S., Lu, C. H., & Chien, K. L. (2012). Effect of calcium based catalyst on production of synthesis gas in gasification of waste bamboo chopsticks. International journal of hydrogen energy, 37(18), 13737–13745.

7. Sridhar, G., Paul, P. J., & Mukunda, H. S. (2001). Biomass derived producer gas as a reciprocating engine fuel—an experimental analysis. Biomass and Bioenergy, 21(1), 61–72.

8. Zainal, Z. A., Ali, R., Quadir, G. A., & Seetharamu, K. N. (2002). Experimental investigation of a downdraft biomass gasifier. Biomass and Energy, 23(4), 283–289.

9. Jayah, T. H., Aye, L., Fuller, R. J., & Stewart, D. F. (2003). Computer simulation of a downdraft wood gasifier for tea drying. Biomass and Bioenergy, 25(4), 459–469.

10. Uma, R., Kandpalb, T. C., & Kishorea, V. V. V. (2004). Emission characteristics of an electricity generation system in diesel alone and dual fuel modes. Biomass and Bioenergy, 27(2), 195–203.

11. Kennedy, Z. R., & Lukose, T. P. (2006). Performance evaluation of a double-walled downdraft gasifier for energy applications. Advances in Energy Research, 307–317.

12. Bhoi, P. R., & Channiwala, S. A. (2009). Emission characteristic and axial flame temperature distribution of producer gas fired premixed burner. Biomass and Bio-Energy, 33, 469–477.

13. Sharma, A. K. (2009). Experimental study on 75 kW, downdraft (biomass) gasifier. Renewable Energy, 34(7), 1726–1733.

14. Wankhade, P. P., Jajoo, B. N., & Wankhade, P. R. (2009). Performance evaluation of parameters for biomass gasifier-diesel engine setup using woody biomass (Subabool). Second International Conference on Emerging Trends in Engineering and Technology, IEEE, 25–29.

15. Iqbal, M., Zainal, Z. A., Mahadzir, M. M., & Suhaimi, H. (2010). Wood gas from the suction gasifier—an experimental analysis. Asian Journal of Applied Sciences, 3(1), 52–59.

16. Kalbande, S. R., Deshmukh, M. M., Wakudkar, H. M., & Wasu, G. (2010). Evaluation of gasifier based power generation system using different woody biomass. ARPN Journal of Engineering and Applied Sciences, 5(11), 82–88.

17. Chawdhury, M. A., & Mahkamov, K. (2011). Development of a small downdraft biomass gasifier for developing countries. J. Sci. Res., 3(1), 51–64.

18. Panwar, N. L., Salvi, B. L., & Reddy, V. S. (2011). Performance evaluation of producer gas burner for industrial application. Biomass and Bioenergy, 35(3), 1373–1377.

19. Patil, K., Bhoi, P., Huhnke, R., & Bellmer, D. (2011). Biomass downdraft gasifier with internal cyclonic combustion chamber: Design, construction, and experimental results. Biomass technology, 102(10), 6286–6290.

20. Centeno, F., Mahkamov, K., Lora, E. E. S., & Andrade, R. V. (2012). Theoretical and experimental investigations of a downdraft biomass gasifier-spark ignition engine power system. Renewable Energy, 37, 97–108.

21. Rajvansi, A. K. (1986). Alternative Energy in Agriculture, Volume II, Ed. D. Yogi Goswami, CRC Press, pp. 83–102.

CHAPTER 13

POTENTIAL OF RENEWABLE SOURCES FROM SOLID BIOMASS IN SERBIAN AND HUNGARIAN AGRICULTURE

GORAN TOPISIROVIC and LASZLO MAGO

CONTENTS

13.1   INTRODUCTION

Renewable energy sources are strongly emphasized among the other items for renewable energy production and environmental protection. Besides, important issues are improvements in rural development, employment, energy supply diversification, lower fossil fuels consumption, reliability of energy supply, and engagement of domestic industry, etc.

According to the EU Directive 2003/30/EC, the biomass is defined as, *"biomass means the biodegradable fraction of products, waste and residues from agriculture (including vegetal and animal substances), forestry and related industries, as well as the biodegradable fraction of industrial and municipal waste."*

In the action plan for biomass 2010–2012 by the Ministry of Environment and Spatial Planning and the Ministry of Energy and Mining of the Serbian Government, this definition is more precise: *"biomass is the biodegradable fraction of products, waste and residues from agriculture (including vegetal and animal substances), forestry and wood industry, as well as the biodegradable fractions of industrial and municipal waste, which use in energy production is allowed, according to the relevant regulation from the field of environmental protection."*

Basically, biomass is a plant or animal matter (hence organic resources) that can be used to produce energy through different processes. The energy of plant matter is recaptured by the plants in the photosynthesis, transforming the sunlight into chemical energy and providing the base for the environmental chain. During the photosynthesis, plants combine carbon dioxide from the air and water from the ground to generate carbohydrates, which form the building blocks of biomass. In this way, the solar energy is stored in the chemical bonds of the structural components of biomass. This energy can be extracted using different methods. On the other hand, the main source of energy from animal sources mainly comes from cattle manure.

This Chapter discusses potential of renewable sources from solid biomass in Serbian and Hungarian agriculture. The general importance of the renewable energy sources in the EU economy is illustrated in Table 13.1.

13.1.1 CROP SPECIES FOR ENERGY

An energy crop is a plant grown as a low cost and low maintenance harvest used to make biofuels, or combusted for its energy content to generate electricity or heat. Energy crops are generally categorized as woody or herbaceous (grassy). Figure 13.1 shows different possibilities for use of field crops [2]. The most common energy crops are: rapeseed, soyabean, jatropha, mahua, mustard, flax, sunflower, palm oil, hemp used for oil and energy (biodiesel), woody crops such as willow or poplar, as well as

TABLE 13.1 Recent and Predicted Renewable Energy Sources Potentials in the EU

Type of energy	Recent renewable energy source potentials	Predicted potentials	Increment index
1. Wind	2.5 GW	40 GW	16
2. Water	92 GW	105 GW	1.1
• Large power plants	82.5 GW	91 GW	0.01
• Small power plants	9.5 GW	14 GW	1.5
3. Photovoltaic cells	0.03 GW	3 GW	100
4. Biomass	44.8 M TOE	135 M TOE	3
5. Geothermal			
• Electricity	0.5 GW	1 GW	2
• Heat	1.3 GWt	5 GWt	3.8
6. Solar collectors	6.5 x 10⁶ m²	100 x 10⁶ m²	15.3
7. Passive solar energy		35 MTOE	-
8. Other		1 GW	-

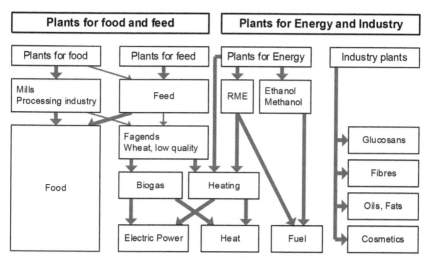

FIGURE 13.1 Plants for energy usage [2].

temperate grasses such as miscanthus and *Pennisetum purpureum* (both known as elephant grass). If carbohydrate content is desired for the production of biogas maize, Sudan grass, millet, white clover and many others, can be made into silage and then converted into biogas. Ethanol can be made from sugar or starchy crops (sugar cane, sugar beet, potato, maize).

13.2   HUNGARIAN PREVIEW

According to surveys, there is a significant biomass potential in Hungary. The total bulk of biomass in the country is up to 350–360 million tons out of which 105–110 million tons (about 30 %) reproduce themselves annually. The energy content of the biomass developing annually is up to 1185 PJ, which is 5 % more than the total annual energy consumption of the country (1120 PJ). The fact that quantity of coal generated annually by plants is four times as much as the quantity of fossil coal exploited for energetic purposes in a year – as much as 30.4 million tons. Potential and utilization possibilities of energetic biomass from the agriculture are detailed in Table 13.2.

TABLE 13.2   Potential and Utilization Possibilities of Energetic Biomass from the Agriculture

		Quantity 1000 t/year		Energy content PJ/year	
No.	Biomass	Min.	Max.	Min.	Max.
I. Biomass for combustion					
1.	Straw	1,000	1,200	11.7	14.0
2.	Stalk	2,000	2,500	24.0	30.0
3.	Energy grass	500	600	6.0	7.0
4.	Vine- and orchard shoots	300	350	4.3	5.0
5.	Energy plants on arable land	1,800	2,500	27.3	38.0
II. Production of biofuels					
1.	Corn/maize	1,200	2,000	14.4	24.0
2.	Wheat/rye	600	1,800	7.2	21.6
3.	Rape	220	460	3.3	7.0
4.	Sunflower	50	200	0.8	3.2
III. Biogas production					
1.	Liquid manure, organic waste	6,000	10,000	5.4	9.0
2.	Silo maize, sorghum	1,600	3,200	5.4	10.8
Total				**109.8**	**169.6**
In % of the total Hungarian energy consumption of 1120 PJ				9.7 %	15.0 %

13.2.1   BIOMASS FOR ENERGY

In the primary biomass produced by the agriculture, first of all the by-products arising in better amount can be reckoned with for energetic purposes. Under common or regular conditions 2.6–2.9 million tons of cereal straw is processed annually of which 1.6–1.7 million tons are utilized for animal breeding and for industrial purposes. The major part of the remaining 1.0–1.2 million tons of cereal straw could be used for energy production and annually 11.7–14 PJ energy could be produced from it. At present, straw is practically not utilized for energetic purpose in Hungary due to the lack of appropriate technology.

Maize stalk production in Hungary is 8–10 million tons of which 2–2.5 million tons could be utilized for energetic purpose which could yield 20–24 PJ energy/year. Among the by-products of crops, sunflower stalk and rape straw also arise in big quantities which could be utilized for burning and could supply 5–6 PJ thermal energy annually should the appropriate technologies for harvesting and burning be available.

The quantity of vineyard and orchard pruning residues (branch tendrils and fruit tree loppings) arising annually is 300–350 thousand tons which could supply 4.3–5 PJ energy. There have only been attempts for their burning till now. The harvesting in bales and burning of branch tendrils is a viable solution on the vine growing farms. For the chopping, collecting and burning of pruning residues, no technology has been developed so far in this country.

Among the plants which can be produced on large farms for energetic purpose first of all the energy-grass and the energetic tree plantations can be considered in Hungary.

The energy-grass as a short rotation herbaceous grass that is able to provide a dry bulk of 10 tons/ha which can be baled for several years with the energy content of 110–120 GJ/ha. The energy-grass can easily be pelletized. The 6–7 tons/ha of pellets can be produced from the grass, the burning features of which are more auspicious in low capacity stokes than that of the chopped material in thermal power stations.

If the final form of firing technology of energy-grass can be developed, then cropping could be started in a short time on 50,000–60,000 hectares that can supply 500–600 thousand tons of bulk of biomass annually, of which 6–7 PJ energy can be produced.

Another prospective source of bio-energy is the energetic tree planta-tion classified in the agricultural plantation management cultivation sector by which dendrimers can be produced relatively fast and in big quantity for energetic purpose.

According to experiences hitherto, it is expedient to plant the short rotation wooden crops varieties (poplar, willow) @ of 12,000–15,000/ha plants which will be ready for felling in 3–5 years. The re-shooting tree stock can be harvested in another 3–5 years by felling totally 5–7 times assuming a plantation lifespan of 15–25 years. On the basis of long term-experiments made with different tree varieties, yields of 11–20 t/ha/year can be achieved, from which 185–330 GJ/ha energy can be produced.

A rapid territorial expansion of the energetic plantation is expected in the near future which can achieve, or even exceed 100 thousand hectares of which 25–30 PJ energy can be gained.

For energy production under arable land conditions, triticale in the form of whole plant cut into windrow and baled can also be taken into account the yield of which may reach 8–10 t/ha with 40 % grain bulk in it. Its energy content is 15–16 GJ/t so that 120–160 GJ/ha energy can be produced. It has a favorable feature from the point of view of firing tech-nology, because in baled form it burns more slowly and producing more heat evenly than wheat straw. These biomasses originating from plants, which can be produced on the field and utilized by direct burning, are gaining a growing emphasis in our national energy policy in the coming years (Table 13.3).

13.3   SERBIAN AGRICULTURAL PREVIEW

The agricultural biomass wastes are coming from cereals, mostly wheat, barley and corn, and from industrial crops mostly sunflower, soya, and rapeseed. In addition, there are many livestock farms in agricul-tural regions, where liquid and solid manures are considered as biomass waste. Fruit growing is also present in the agricultural areas, but the main area of fruit growing is the hilly region in the south, where main crops are plums, apples, cherries, peaches, and grapes. Actual annual biomass production in Serbia is about 12.5 million tons (2.7 million of

TABLE 13.3   The Real and Feasible Capacity for Energetic Utilization of Solid Biomass in Hungary

Biomass	Utilization	Actual capacity			Expected growth till 2020		
		Unit (Nos.)	Capac-ity (MW)	Biomass demand (2,000 t)	Unit (Nos.)	Capac-ity (MW)	Biomass demand (1,000 t)
Wood chips (forest or planted wood)	Electricity	5	140	1,000	8	420	2,800
	Central heating	5	24	25	25	120	150
	Central heating + electric energy production	2	12	32	20	120	180
Straw, energy grass	Straw power plant, electric energy production and heat utilization	-	-	-	2–3	40–60	450
Total		12	176	1,057	55–56	700–720	~3,600

TOE). Out of this, 1.7 million TOE is agricultural biomass, and 1.02 million TOE comes from the forestry (Table 13.4). Figure 13.2 indicates regions of Siberia under corn and plum production.

For easier classification, biomass that originates from agriculture can be divided in three main categories: from crop production, fruit production and livestock breeding.

13.3.1   BIOMASS FROM CROP PRODUCTION

In Serbia, there are many small individual land owners who deal with production of cereals or industrial plants, like sunflower or soya. A great deal of crop farming production, almost 75% is achieved in small or medium size private ownership, while only about 25% of crop farming production belongs to agricultural companies of relatively larger size (Table 13.5).

TABLE 13.4   Possibilities for Energy Production From Biomass in Serbia

Biomass source	Potential (TOE)
Wood biomass	1,527,678
Wood for combustion	1,150,000
Wood residues	163,760
Wood processing residues	179,563
Outside forests wood	34,355
Agricultural biomass	1,670,240
Crop production residues	1,023,000
Fruit and grape production and processing residues	605,000
Liquid manure (for biogas production)	42,240

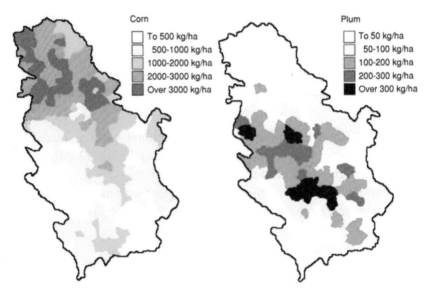

FIGURE 13.2   Agricultural areas under corn fields and plum orchards.

The modern way of livestock breeding does not consider extensive use of biomass residues for animal bedding. At large agricultural farms, it is more favorable and cost effective to collect biomass residues in bales, and use them without any further preparation in small or medium sized boilers.

TABLE 13.5 Yield of Main Species in Crop Farming and Energy Potential of Their Residues

Plant	Yield (10³ t)	Total residues (10³ t)	Residues for energy use (10³ t)	Energy potential (TOE)
Wheat	29,050	29,050	13,650	
Barley	3,650	2,950	1,800	
Rye	141	155	44	Average heating value 14 MJ/kg
Corn	48,270	53,100	11,400	
Sunflower	2,800	7,050	2,400	
Soya	1,600	3,200	1,300	
Rape seed	26	78	16	
Total		95,600	30,600	1,023,000

About half of biomass residues at large agricultural farms can be used for energy purposes, while only about 20% biomass residues generated on relatively small private farms can be used for energy purposes.

Greater amount of biomass residues generated on small agricultural farms can be used for energy if these owners would have appropriate ovens and boilers for burning biomass residues, or if they find an interest to collect residues and sell them.

13.3.2 BIOMASS FROM FRUIT PRODUCTION AND VITICULTURE

One of the main activities in fruit growing and viniculture is pruning of small branches, and these cut small branches can be available for energy purpose. Total number of registered fruit trees is about 94 x 10⁶. The 50% of this are plum trees, about 20% are apple trees and almost 15% are cherry trees, both sour and sweet cherry.

The total biomass residues from fruit growing amounts about 475,000 tons. With average heating value of 14 MJ/kg, the energy potential of biomass residues from fruit trees pruning is about 159,000 TOE. The energy potential of vine pruning residues is about 155,000 TOE (Table 13.6).

Stones of plums, cherries, peaches, and apricots together with peels and seeds of apples, pears, and grapes are wastes derived from processing of

TABLE 13.6   Energy Potential of Biomass Residues Deriving From Fruit Cultivation and Processing

Species	Number of trees [10³ ha]	Type of biomass residues	Biomass residues [t]	Annual energy equivalent [TOE]
Plum	50,630	Pruning, stones	393,500	132,600
Apple	17,570	Pruning, peel	36,200	10,900
Cherries	12,280	Pruning, stones	55,000	16,500
Pear	7,080	Pruning, peel	14,000	4,300
Peach	4,450	Pruning, stones	35,100	11,700
Apricot	1,900	Pruning, stones	15,500	4,100
Walnuts	2,100	Pruning, shell	55,000	14,100
Grape	77,390	Pruning, peel, seeds	515,000	166,300
Total	*360,500*			

fruits. The quantity of these wastes amounts to about 200,000 tons. With a relatively modest heating value of 9 GJ/t, the energy potential of fruit processing wastes is about 46,000 TOE. This value is relatively small compared to the energy potential of other fruit residues derived from growing. But an important advantage of these wastes is that, they are already collected in every company dealing with fruit processing.

The overall energy potential of biomass residues from fruit growing, viniculture and fruit processing is about 605,000 TOE.

13.3.3   BIOMASS FROM LIVESTOCK BREEDING

Liquid manure deriving from cattle and pig breeding together with poultry litter are potential energy sources as well. Because of high water content (up to 90%), these slurries are usually treated by anaerobic digestion. These wastes are recommended for anaerobic digestion, not only for an energy reason, but also for getting more suitable and environmentally friendly fertilizers. Livestock breeding in Serbia comprises of mainly cattle, pigs, poultry and sheep.

The major part of livestock is located in small farms, with only a few heads in each. An organized manure collection from these small farms

TABLE 13.7   Livestock in Medium and Great Farms and Energy Potential of Their Manure [1, 7]

Live-stock	Location of farms	Number of heads	Manure [m³/day]	Biogas [m³/day]	Annual energy equivalent [TOE]
Cattle	Flat regions	149,300			
	Hilly regions	111,000			
	Total	260,300	5,270	105,000	20,140
Pigs	Flat regions	1,369,500			
	Hilly regions	285,600			
	Total	1,655,100	4,560	91,200	17,500
Poultry		2,350,000	480	24,000	4,600
Total					42,240

is not likely to be easily technically feasible, and the financial feasibility is uncertain as well. Therefore, in the analysis of energy potential, only manure in medium and great farms is considered as a prospective source of fuel, since manure from these farms does not need to be transported, and can be efficiently treated in an aerobic digestion (Table 13.7).

13.4   COMPETITION BETWEEN FOOD AND ENERGY PRODUCTION

FAO and OECD estimate that food consumption will increase 10% annually, while energy consumption will increase 3% by 2030. Increased need of food will be covered with bio-technical progress before all on suitable location for agricultural production. Some analysts note that oil prices will not stay on present price of US$87.63 a barrel (the highest level since late 2008). Considering that oil price has exceptional influence on bioenergy production, production of bioethanol in Brazil is rentable without subsidies if the oil price is between US$30 and US$40 per barrel. In such areas more agricultural land will be used for energy production and higher food prices will be reasonable consequence.

EU commission evaluated profitability of biofuel production in Europe when oil prices are between US$60 and US$90 a barrel which means double than breakeven point in Brazil. According to Schenkel [8], food production in Europe will be still in the foreground ahead energy production. According to Henniges et al. [4] calculations for EU countries, domestically produced biofuels would not be viable without a subsidy of some kind unless oil prices were consistently higher than US$80 a barrel. Given that such prices are not imminent, the biofuel industry in Europe, as in the United States, is heavily dependent on continuing political support. The European Union has supported biofuel production primarily to promote sustainable farming, protect the countryside, create additional value added and employment in rural areas, reduce the cost of farm support policies, and diversify its energy supplies. Reducing emissions of greenhouse gases is only a secondary goal because the net energy efficiency of the biofuel crops grown in Europe is low. Thus the biofuel industry has much higher carbon abatement costs than do some other fields of energy use.

Partly as a result of negative publicity regarding biofuels, the European Union watered down its 2020 biofuels conversion goals while Germany began to remove tax credits that aided its domestic biodiesel industry. The biofuel tax increases, aimed at ultimately creating tax parity between biofuels and conventional fuels, rendered the domestic German biodiesel industry unable to compete with subsidized biodiesel from South America and the US. The 27% of German capacity shut down altogether, while 36% ran at less than 50% of capacity. Meanwhile, the European Union flirted with doing away with a 10% biofuels target and 2008–2020 conversion schedule. The EU ultimately agreed to confirm the targets as a renewable energy conversion, but 30% of the target would be met by electric cars or trains, with the remainder to come from biofuels. The EU also said it would develop regulations to limit the impact of indirect land-use change, while biofuels developed from non-food sources will receive preferred treatment under the agreement. The agreement will need to be ratified by the European Parliament and all 27 EU members.

13.4.1 REAL MEANING OF BIOENERGY AND REALITY OF OTHER SOURCES

Actual results of bioenergy use are not much encouraging except direct burning of biomass. Bioenergy contribution is marginal in total energy balance. If we use half of all arable land for bioenergy production, it will cover only 5% of energy needs. Although some research is in progress, there is no new kind of bioenergy on market such as BtL (Biomass-to-liquid) [3]. There are prospects that mankind will have to secure and cheap sources of energy such as nuclear fission and fusion (atomic power plant for the future) [5, 6].

13.5 CONCLUSIONS

The energy crisis on the world draws the attention to the energy sources which can be produced by the agriculture. The lasting energy deficiency can be replaced with the big mass of biomass gained mostly from the agriculture and forestry. The agriculture would be capable to cover 10% of the country's energy needs from renewable energy sources on a middle term. A new power generating section of the agriculture takes shape across Europe in the immediate future expectedly, that may contribute in a considerable measure to the reduction of the energy deficiency collaborating tightly with the energy producer's and the service provider's sections of the countries, while he secures new revenue source.

- Bioenergy contribution is marginal in total energy balance except direct burning of biomass.
- Economy of bioenergy production depends of subsidies in EU and without them only biomass combustion is cost effectively.
- Producer's dilemma what is better to produce: food or energy depends of income and household capacities.
- Ecological acceptability of bioenergy is not always positive.
- Ethnical questions (burning of grains) sometimes are significant but economy is in first plan.
- Plant use for energy in EU will not significantly decrease food production but it will increase food prices.

13.6 SUMMARY

The research was aimed to measure the quantity of agricultural biomass suitable for energy purposes at regional level. Furthermore, this research also aimed: to determine the potential of biomass for energy purposes with regards to the grown plants; and to name the possibilities and ways of utilization of the solid biomasses of various origins.

Experiments of this kind have already commenced in Hungary and Serbia. The potentials of different types of solid biomass from agriculture are presented in this chapter. The survey has included the comparative presentation of solid biomass potentials in Hungary and Serbia.

KEYWORDS

- agriculture
- energy
- Hungary
- potential
- renewable energy
- Serbia
- solid biomass
- sources

REFERENCES

1. Anonymous (2010). Akcioni plan za biomasu: 2010–2012. Ministarstvo rudarstva i energetike Republike Srbije, Vlada Republike Srbije. Beograd.
2. Breitschuh, G., Reinhold, G., & Vetter, A. (2005). Wirtschaftlichen bedeutung der energetischen nutzung nachwachsender rohrstoffe für landwirtschaft: Der Landwirt als Energiewirt. Potenziale für die Erzeugung. KTBL-Schrift, 420, 19–36.
3. FNR (Fachagentur Nachwachsende Rohrstoffe), (2006). BtL: Biokraftstoff der Zukunft. Landtechnik, 4, 206–207.
4. Henniges, O., & Zeddies, J. (2005). Economics of Bioethanol in the Asia-Pacific: Australia – Thailand – China. In: F. O. Licht's World Ethanol and Biofuels Report, 3(11).
5. Oljača, S. (2007). Čiste tehnologije i očuvanje životne sredine u poljoprivredi. Poljoprivredni calendar, p. 323.
6. Potočnik, J. (2006). Zlitje prihodnosti. National Geographic, September, 31–38.
7. Quirin, M., & Reinhart, G. A. (2005). Ökobilanzen von Bioenergieträgern – ein Überblick. KTBL-Schrift, 420, 37–45.
8. Schenkel, R. (2006). Z dejstvi nad predsodke in strahove. Delo, July 20, 17.

CHAPTER 14

BIODIESEL PRODUCTION: A REVIEW ON *SWIDA WILSONIANA* (WILSON'S DOGWOOD)

PEIWANG LI, YOUPING SUN, CHANGZHU LI, LIJUAN JIANG, LIANGBO ZHANG, and ZHIHONG XIAO

CONTENTS

14.1 INTRODUCTION

Biodiesel as an alternative to petrodiesel has recently drawn considerable attention due to diminishing availability of discoverable fossil fuel reserves and environmental consequences of exhaust gases from fossil fuels. Biodiesel, alkyl esters of fatty acids, is made from renewable resources including animal fats and vegetable oils. The edible fatty oils derived from coconut, linseed, palm, rapeseed, soybean, and sunflower have been exploited commercially [10]. Therefore, biodiesel feedstock may compete with food supply and/or food crops for arable land in the long term. Production of biodiesel from such edible oils is not feasible in China since there is a huge population of 1.35 billion and relatively inadequate arable land resources per capita [11, 12].

Oil bearing plants that produce non-edible oils in appreciable quantity and adapt to non-cropped marginal lands and wastelands would be better feedstock of choice for biodiesel production. Fortunately, there are more than 4,000 species of plants in China with potential for bioenergy production [20]. A total of 38 oilseed crops have been identified as potential energy plants in China and are mainly distributed in tropical and subtropical areas [17, 19]. *Swida wilsoniana* (Wangerin) Soják (syn. *Cornus wilsoniana*; Wilson's dogwood or guangpishu) was proposed as a candidate energy plant [11, 12, 14].

Wilson's dogwood, a member of *Cornaceae*, is a deciduous tree that can grow up to 8–10 m tall and produces non-edible oil [7]. It has green leaves, cream white flowers, purple brown fruits, and beautiful exfoliate barks (Figure 14.1). It is native to forest lands in China with an elevation of 100–1000 m, an average temperature of 18–25°C, and an annual precipitation of 1000–1570 mm [17]. It is a potential bioenergy feedstock because it is a fast-growing and easy-to-harvest plant with high oil content in fruit and high adaptability to marginal land [17, 19]. It can maintain maximum productivity for over 50 years with an average of 50 kg dry fruit per year produced on a mature tree. The fruit oil content can reach up to 33~36% [27]. During the period of 1997–2007, six high-yielding and stress-tolerant Wilson's dogwood cultivars ('Xianglin G1,' 'Xianglin G2,' 'Xianglin G3,' 'Xianglin G4,' 'Xianglin G5,' 'Xianglin G6': China) were selected using conventional breeding methods such as mass selection, elite-tree selection, and clone testing [14, 15, 17]. They are important breeding materials for the development and improvement of this plant for biodiesel usage.

Biodiesel production from Wilson's dogwood fruit oil at bench scale has been investigated for ten years by Hunan Academy of Forestry. Recently, this process was scaled up, and about 3000 tons of biodiesel is produced from Wilson's dogwood fruit oil every year.

In this chapter, authors have reviewed research and summarized the progress on the propagation of Wilson's dogwood and the protocol of biodiesel production from Wilson's dogwood fruit oil.

FIGURE 14.1 Wilson's dogwood plants with green leaves, cream-white flowers, purple brown fruits, and exfoliated bark (inset).

14.2 PROPAGATION OF SWIDA WILSONIANA

14.2.1 SEED PROPAGATION

Stratification was required for breaking the double dormancy of Wilson's dogwood seeds [15], similar to most dogwood seeds such as *C. alternifolia* L., *C. controversa* Hemsl., *C. florida* L., *C. kousa* (Buerger ex Miq.) Hance, *C. mas* L., *C. officinalis* Sieb. & Zucc. [23]. A cold treatment of 2–3 months at 2–4°C (35–40°F) was required before sowing [2]. Liu et al. [21] recently documented that 82% seed germination was achieved by immersing *C. florida* seeds in 98% sulfuric acid for 10 mins, then soaking the seeds in 500 mg·L^{-1} gibberellic acid (GA$_3$) for 72 hours before cold stratification at 5°C for 60 days. Alternatively, Li et al. [15] reported that seeds were soaked for 2 hours in 70°C NaHCO$_3$ solution (1%, w/v), surface disinfected for 24 hours with 0.1% KMnO$_4$, and treated for 2 hours with 10% H$_2$SO$_4$. About 68.4% of treated seed germinated under condition of 25°C and 80–85% relative humidity. This protocol resulted in a higher percentage of germination without a cold pretreatment. The seeds remain viable for only two years. Not surprisingly, seed cultivation of Wilson's dogwood is limited because seed viability rapidly decreases and seedling variation occurs. Moreover, seedlings often take 7–8 years to attain maturity and flower [11].

14.2.2 CUTTING PROPAGATION

Dogwoods are usually propagated from cuttings. Raulston [23] documented that softwood and semi-hardwood cuttings collected from May to August were suitable for the species, cultivars, or hybrid of *C. alba* L., *C. controversa*, *C. florida*, *C. kousa*, *C. mas*, *C. officinalis*, *C. sanguineas* L., and *C. sericea* L., while hardwood cuttings collected in November to January were good for *C. alba*, *C. sanguineas*, and *C. sericea*. Similar to aforementioned dogwoods, Wilson's dogwood can be propagated through softwood and hardwood cuttings. Semi-hardwood cuttings were treated with 1000 mg·L^{-1} 3-indole butyric acid (IBA) using quick dip and inserted in a carbendazim-sterilized substrates (1 peat moss: 1 perlite: 1 red soil, v/v/v) [17]. Roots began to form at 25 days of incubation, and 93.4% of cuttings formed roots with 11.8 roots per cutting and 10.9 cm long. Analysis of plant growth regulators in Wilson's dogwood cuttings showed that a high concentration of endogenous indole-3-acetic acid (IAA) and low concentration of endogenous abscisic acid (ABA) improved the formation and development of root primordia in cuttings, while a high level of endogenous GA$_3$ in cuttings inhibited adventitious root formation. Endogenous zeatin (4-hydroxy-3-methyltrans-2-butenylaminopurine) stimulated the differentiation and formation of root primordia of Wilson's dogwood cuttings at low concentrations, while it promoted the growth and development of root primordia at higher concentrations. In addition, initial concentrations of IAA and zeatin have a positive correlation with the rooting percentage, while initial ABA and GA$_3$ concentrations negatively correlated with rooting percentage.

14.2.3 GRAFTING

Grafting is among the most expensive propagation technique. It is used only for superior and high value crops. Grafting such as chip-bud and t-bud are good techniques to propagate most important dogwood cultivars or hybrids from July to September [23]. An ideal rootstock is needed for the success of grafting. The same *Cornus* species are usually chosen as the rootstock for grafting the corresponding cultivars or hybrids [23]. In some cases, other species could be used, for instance, *C. florida* is an ideal rootstock for *C. kousa* cultivars [23]. In the early stage of the project (e.g., clone testing) at Hunan Academy of Forestry, graftage is the most feasible way to propagate many plants for phenology, morphology and yield observation.

Li et al. [19] reported that 0.15–0.2 cm buds of Wilson's dogwood could be successfully grafted using t-budding on one-year-old seedlings. About 80% of grafted plants survived. The grafted plants matured and flowered after 2–3 years of growth [11], which is significantly less than the time to flowering for seedlings. Now graftage continues to be used for improving plant quality, yield, dwarf forms, and better adaptation to ecological ranges. The efforts to breed dwarf plants have achieved great success. Five years after transplanting, the plant is less than 3 m tall, and the crown is reduced by 60% (Figure 14.2). This plant also begins to set fruit three years after transplanting.

14.2.4 TISSUE CULTURE

Micropropagation was successfully established for *C. mas* [3, 4], *C. florida* [9, 24], *C. nuttallii* Aud. [5], and *C. Canadensis* L. [6]. Authors developed a

FIGURE 14.2   Wilson's dogwood plant in the wild (left) and a dwarf plant produced through graftage (right).

micropropagation protocol for propagating the elite trees of Wilson's dog-wood. A single-node segment of Wilson's dogwoods collected in autumn were sterilized in 75% alcohol for ten seconds dip and then in 0.1% mercury for eight minutes. These explants were then cultured on Murashige-Skoog (MS) medium [22] supplemented with N^6-benzyladenine (6-BA) to induce the emergence of axillary shoots that served as explants for multiplication and rooting. The most suitable medium for shoot induction was MS medium with 2.0 mg·L^{-1} 6-BA, 0.05 mg·L^{-1} **α-naphthaleneacetic acid** (NAA), 300 mg·L^{-1} polyvinylpyrrolidone (PVP), and 3% sucrose. Regenerated shoots could be proliferated on woody plant medium (WPM) plus 0.5 mg·L^{-1} 6-BA, 0.5 mg·L^{-1} zeatin, 0.1 mg·L^{-1} IBA, 300 mg·L^{-1} PVP, and 5.5 shoots per explant formed. Shoots were quick-dipped in 1.0 g· L^{-1} IBA potassium solution for five seconds and cultured on ½ **strength WPM containing** 300 mg· L^{-1} PVP to induce roots. About 95.2% of shoots formed roots with 4.7 roots per shoot and 2.8 cm long.

Plantlets with well-developed roots were transplanted to small plastic pots containing a mixture of perlite and peat (3:1, v:v). The roots were thoroughly washed in distilled water before transplant. Plants were covered with transparent plastic and maintained under mist system in a glasshouse with approximately 90% relative humidity, $25 \pm 2°C$ temperatures, over an 11–12 hour photoperiod. One month later, the plastic cover was removed. Ten days later, plants were moved out of the glasshouse and planted in the field. The survival rate was over 95% one month after transplanting. Recently, Li et al. [18] reported an improved micropropagation protocol of Wilson's dogwood through shoot culture. Shoot explants were cultured on Driver and Kuniyuki Walnut medium (DKW) supplemented with 0.5 mg·L^{-1} zeatin, 0.1 mg·L^{-1} IBA, and 300 mg·L^{-1} PVP. More than four axillary shoots were produced per explant. These axillary shoots could either be used as explants for additional shoot production or be rooted in DKW medium containing 1.0 mg·L^{-1} KIBA and 300 mg·L^{-1} PVP. The produced plantlets were transplanted into a substrate (20% clay soil, 40% carbonized rice hull, 20% perlite, and 20% coarse sand) and grew vigorously in a shaded greenhouse under a maximum photosynthetic photon flux density of 200 μmol·m^{-2}·s^{-1}. These established protocols could lead to mass propagation of the high-yield cultivars of Wilson's dogwood. They are also good for breeding horticultural and economically important Wilson's dogwood cultivars with resistance to dogwood anthracnose and powdery mildew.

14.3 PRODUCTION OF BIODIESEL FROM SWIDA WILSONIANA

14.3.1 OIL EXTRACTION FROM WILSON'S DOGWOOD FRUIT

Although cold-pressed oil extraction is the most practical technique for Wilson's dogwood fruit on a large scale, the oil recovery and purity are low [11, 12]. Xiao et al. [26] reported oil extraction from several parts of Wilson's dogwood fruit using petroleum ether. The fruit flesh and seed powder were mixed with ether (1:3 w/v) and incubated in a water bath at 50°C for 6 hours. The oil content of fruit flesh is 67% and seed 33%. Microwave and ultrasound-assisted extraction, and supercritical CO_2 extraction were investigated at a bench scale using an orthogonal experiment design [11]. Mature dark-black Wilson's dogwood fruits were dried under sunlight and ground to a 30- to 40-mesh powder. Microwave assisted extraction (720 w for 7 min) was done with a mixture of petroleum ether and fruit powder (3:1, v/w). The extraction solution was purified using a low-pressure distillation, further washed with Na_2SO_4, and dried for 24 h. The oil yield was 28.1±0.2%. Ultrasound-assisted extraction was carried out with fruit powder being pretreated with hexane for 18 h at a ratio of 1:7 (w/v) and extracted in an ultrasound extractor with a setting of 40 KHz frequency and 180 W for 15 h. The extraction solution was cooled down to room temperature and transferred to a flask. The extraction solution was purified using a low-pressure distillation, washed with Na_2SO_4, and dried for 24 h. The oil yield was 27.8±0.2%. Supercritical CO_2 extraction was applied by extracting fruit powders for 80 min at the optimum extraction conditions: 37 MPa pressure, 30°C temperatures, and 30 kg·h^{-1} CO_2 flow rate. The oil yield was 34.8±0.3%. Palmitic acid (hexanoic), linoleic (9, 12-octadecenoic) acid, oleic (9-octadecenoic) acid, and stearic (octadecenoic) acid are the major fatty acids of Wilson's dogwood fruit oil from the above three methods. They accounted for 92.4% of the total oil composition. The saturation level is 38.7%. The oil extracted from supercritical CO_2 extraction has higher quality with longer shelf time. Furthermore, addition of

0.02% TBHQ (tert-butylhydroquinone) and 0.005% citric acid inhibited the rancidification of Wilson's dogwood oil and extended the shelf life from 4 to 9 months at 20°C [16].

14.3.2   ACID PRETREATMENT

Since crude fruit oil of Wilson's dogwood has a high acid value, a pre-treatment is required to reduce acid for biodiesel conversion. He et al. [8] reduced the acid value of the crude fruit oil of Wilson's dogwood from 23 mg KOH/g to 1.4 mg KOH/g. Li et al. [14] followed the refinery process as described by He et al. [8] and decreased the acid value of crude fruit oil from 13.2 mg KOH/g to 3.1 mg KOH/g using phosphoric acid and alcohol. Acid reduction of the crude fruit oil of Wilson's dogwood was also carried out by Xiao et al. [26]. The Wilson's dogwood seeds were extracted using the aforementioned supercritical CO_2 to produce crude oil that was mixed with methanol (1:2.4, v/w) at 30°C for 10 min to reduce acid. The acid number of the crude oil decreased from 22.6 mg KOH/g to 1.3 mg KOH/g.

14.3.3   FATTY ACID TRANSESTERIFICATION AND DECARBOXYLATION

The fatty acid methyl ester is produced by mixing Wilson's dogwood oil with methanol (1:5 – 1:7 molar ratio) and sodium methoxide (CH_3ONa) (0.9–1.1 mass ratio) at 60°C for 100 min [12]. The yield of biodiesel was above 75%. The dynamic viscosity was 5.6 $m^2 \cdot s^{-1}$ and density of biodiesel 0.891 $g \cdot cm^{-3}$, while the cetane number was 57°C and flash point >105°C. The fuel properties of the biodiesel made from Wilson's dogwood oil are similar to the 0[#] petroleum diesel (Table 14.1). Li et al. [13] developed an improved protocol to convert Wilson's dogwood oil to fatty acid methyl ester using 1-butyl-3-methylimidazolium hydroxide ([Bmim]OH), an ionic liquid catalyst with high activity and stability. Wilson's dogwood oil was mixed with methanol at a molar ratio of 6:1 and catalyzed with 0.9% [Bmim]OH at 60°C for 70 min. The yield of fatty acid methyl ester was 96.4%.

TABLE 14.1 Comparison of the Biodiesel Produced From Wilson's Dogwood and 0# Petroleum Diesel Fuel [13]

Item	Analytical protocol	Biodiesel from dog-wood oil	GB/ T 20828	0# Petroleum diesel
Density at 20°C (g·cm⁻³)	GB/T 2540	0.876	0.82–0.90	NA
Dynamic viscosity at 40°C (mm²·s⁻¹)	GB/T 265	4.2	1.9–6.0	3.0–8.0 at 20°C
Flashpoint (Pensky Martens Closed Tester, PM, °C)	GB/T 261	170	≥ 130	≥ 55
Cold-filtering plugging point (°C)	SH/T 0248	0	NA	> 4
Carbon residue on 10% distillation residue (%)	GB/T 17144	0.1	< 0.3	< 0.3
Cetane number	GB/T 386	49	≥ 47	≥ 49
Copper corrosion rating (50°C, 3 h)	GB/T 5096	1a	1a	< 1
Acidity (mg KOH·g⁻¹)	GB/T 264	0.15	≤ 0.8	< 7
Distillation temperature (°C) at 90% vol. recovered	GB/T 6536	340	360	< 355
Color number	GB/T 260–1986	1.0	-	≤ 3.5
Water (% by vol.)	SH/T 0246	0.001	≤ 0.05	trace
Sulfur (mg·kg⁻¹)	SH/T 0689	0.002	≤ 500	≤ 350
Ash (% by vol.)	GB/T 2433	0.008	≤ 0.02	< 0.01
Glycerol (% by vol.)	ASTM D6584	0.11	≤ 0.24	-
Oxidation stability [mg·(100 mL)⁻¹]	SH/T 0175	0.6	6.0 (110°C /h)	< 2.5

Ding et al. [1] reported transesterification of Wilson's dogwood oil with methanol to produce biodiesel using an immobilized lipase Lipozyme TL IM as the catalyst in a $MgCl_2$-saturated solution. The yield of biodiesel reached up to 86.5% after reaction for 8 hours under the optimal conditions: 20% of Lipozyme TL IM based on Wilson's dogwood oil, methanol/oil molar ratio of 3, and agitation speed of 150 r·min⁻¹. This study resolved the problem of immobilized lipase inactivation using the $MgCl_2$-saturated solution.

Hydrocarbon fuel by microwave-assisted pyrolysis decarboxylation was produced using sodium soap made from Wilson's dogwood fruit oil [25]. Both the yield of the liquid sodium soap pyrolysis product and the hydrocarbon content of the liquid product were usually above 70%. The dynamic viscosity is 2.09–2.85 $m^2 \cdot s^{-1}$ and density 0.850–0.875 $g \cdot cm^{-3}$, similar to petro-diesel (Table 14.1).

14.4 CONCLUSIONS

Non-edible oil woody plants would be feasible biodiesel feedstocks and play a key role in renewable bioenergy resources. Production of biodiesel from the fruit oil of Wilson's dogwood is beneficial not only for energy security, bioenergy research, and commercial production, but also for environmental protection. Wilson's dogwood is a non-edible oil plant with high oil yield and strong adaptability. It can be cultivated in limestone and non-arable marginal lands. The cost of biodiesel produced from the fruit oil of Wilson's dogwood is not yet determined. However, plant breeding work, cultivation system, clean and efficient conversion technology, and comprehensive utilization technology need further investigation to lower the cost of biodiesel from Wilson's dogwood oil.

14.5 SUMMARY

Biodiesel as an alternative to petro-diesel has recently drawn considerable attentions since the availability of discoverable fossil fuel reserves is limited and the use of fossil fuels causes many environmental problems. Biodiesel feedstock may compete with food supply and/or food crops for arable lands in the long term because it is usually produced from animal fats and vegetable oils. Wilson's dogwood produces non-edible oil and is a feasible feedstock for biodiesel production. Here we summarize the progress of propagation and biodiesel production of Wilson's dogwood in the past years.

ACKNOWLEDGEMENTS

This work was partially supported by National Science and Technology Support Program, Ministry of Science and Technology, China (2015BAD15B02). The content is solely the responsibility of the authors and does not necessarily represent the official views of the funding agency.

KEYWORDS

- biodiesel
- cutting
- Dogwood
- grafting
- micropropagation
- oil extraction
- seed propagation
- tissue culture
- transesterification

REFERENCES

1. Ding, R., Zhong, S. A., Li, N., & Yang, J. J. (2010). Transesterification of Swida wilsoniana oil with methanol to biodiesel catalyzed by Lipozyme TL IM in MgCl2-saturated solution. Journal of Fuel Chemistry and Technology, 38(3), 287–291.
2. Dirr, M. A., & Heuser, C. W. Jr. (1987). The reference manual of woody plant propagation, from seed to tissue culture. Varsity Press, Inc. Athens, GA.
3. Ďurkovič, J. (2008). Micropropagation of mature Cornus mas 'Macrocarpa.' Trees, 22, 597–602.
4. Ďurkovič, J., & Bukovská, J. (2009). Adventitious rooting performance in micro-propagated Cornus mas. Biological Plantarum, 53, 715–718.
5. Edson, J. L., Wenny, D. L., & Leege-Brusven, A. (1994). Micropropagation of pacific dogwood. HortScience, 29, 1355–1356.
6. Feng, C., Qu, R., Zhou, L., Xie, D., & Xiang, Q. (2009). Shoot regeneration of dwarf dogwood (Cornus canadensis L.) and morphological characterization of the regenerated plants. Plant Cell Tiss. Organ Cul., 97, 27–37.

7. Flora of China Editorial Committee. (1994). Flora of China. Science Press, Beijing. p. 56–59.

8. He, H., Yuan, J., He, R., Ren, Z., Yang, H., & Ju, X. (2011). Refining process of Swida wilsoniana oil and its DPPH scavenging activity. China Oils and Fats, 36(5), 36–40.

9. Kaveriappa, K. M., Phillips, L. M., & Trigiano, R. N. (1997). Micropropagation of flowering dogwood (Cornus florida) from seedlings. Plant Cell Reports, 16, 485–489.

10. Korbitz, W. (1999). Biodiesel production in Europe and North America, an encouraging prospect. Renewable energy, 16, 1078–1083.

11. Li, C. (2010). Genetic diversity of plus tree and its fruit fatty acid in Swida wilsoniana. PhD thesis, Beijing Forestry University, Beijing.

12. Li, C., Jiang, L., Li, P., Zhang, L., Xiao, Z., & Li, D. (2005). Production of biodiesel utilizing Cornus wilsoniana fruit oil. Chinese Journal of Bioprocess Engineering, 3(1), 42–44.

13. Li, C., Zhang, A., Xiao, Z., Jiang, Li, & Li, P. (2011). Preparation of biodiesel from Cornus wilsoniana fruit oil with basic ionic liquids as catalyst. Journal of Central South University of Forestry & Technology 31(3), 38–43.

14. Li, C., Liu, Y., Luo, J., Li, R., Yuan, R., & Liu, C. (2007). Study on Cornus wilsoniana oil refining. Cereals and Oils Processing, 11(3), 76–78.

15. Li, P., Li. C., Jiang, L., & Sun, Y. (2006), Seed germination vigor of three species of woody oil plants. Nonwood Forest Research, 24(1), 71–73.

16. Li, R., Liu, Y., Zhou, H., Zeng, W., Han, D., Peng, H., & Ruan, R. (2009). Oxidation stability of oil from Swida wilsoniana. Food Science, 30(21), 87–89.

17. Li, X., Hou, S., Su, M., Yang, M., Shen, S., Jiang, G., Qi, D., Chen, S., & Liu, G. (2010). Major energy plants and their potential for bioenergy development in China. Environmental Management, 46, 579–589.

18. Li, Y., Wang, X., Chen, J., Cai, N., Zeng, H., Qian, Z., & Wang X. (2015). A method for micropropagation of Cornus wilsoniana: an important biofuel plant. Industrial Crops and Products, 76, 49–54.

19. Li, Y., Zeng, H., Wang, X., & Cai, N. (2010). Changes of endogenous hormones during Swida wilsoniana Wanger cuttings. Chinese Agricultural Science Bulletin, 26(15), 247–251.

20. Lin, C., Li, Y., Liu, J., Zhu, W., & Chen, X. (2006). Diversity of energy plant resources and its prospects for the development and application. Henan Agricultural Sciences, 12, 17–23.

21. Liu, H., Qian, C., Zhou, J., Zhang, X., Ma, Q., & Li, S. (2015). Causes and breaking of seed dormancy in flowering dogwood (Cornus florida L.). HortScience, 50(7), 1041–1044.

22. Murashige, T., & Skoog, F. (1962). A revised medium for rapid growth and bioassay with tobacco tissue cultures, Physiologia Plantarum, 15, 473–497.

23. Raulston, J. C. (1996). Propagation Guide for Woody Plants. JC Raulston Arboretum at NC State University, Raleigh, NC. p. 89.

24. Sharma, A. R., Trigiano, R. N., Witte, W. T., & Schwarz, O. J. (2005). In vitro adventitious rooting of Cornus florida microshoots. Scientia Horticulturae, 103, 381–385.

25. Wang, Y., Liu, Y., Yuan, R., Wan, Y., Zhang, J., & Peng, H. (2011). Production of renewable hydrocarbon fuels—Thermochemical behavior of fatty acid soap decarboxylation during microwave-assisted pyrolysis. Materials for Renewable Energy and Environment 2011 International Conference, Shanghai, China, 20–22 May 2012.

26. Xiao, Z., Li, C., Chen, J., Liu, Y., & Wu, X. (2009). Analysis of oil composition in different parts. China Oils and Fats, 34(2), 72–74.

27. Zeng, H., Fang, F., Su, J., Li, C., & Jiang, L. (2005). Extraction of seed oil of Swida wilsoniana by supercritical CO2, microwave and ultrasound. Journal of the Chinese Cereals and Oils, 20(2), 67–70.

PART 6

NANOTECHNOLOGY APPLICATIONS IN AGRICULTURE

CHAPTER 15

NANOTECHNOLOGY AND ITS APPLICATIONS IN AGRICULTURE

RAJNI SINGH, K. P. SINGH, and S. P. SINGH

CONTENTS

15.1 INTRODUCTION

Advances in technology have played a significant role in transforming our agriculture. It has changed us from food gathers and hunters to the efficient and reliable user of the natural resources to wield enhanced control over food, feed and fiber supply. Agriculture is no longer a practice of growing crops and animals for subsistence but it has evolved as an industry over the time. Advances in sciences have had been of immense help in enhancing food production. Amidst this gloomy scenario, food security in developing world science holds a great hope. Research in agriculture has had successes in improving the efficiency of crop production, food processing, food safety, food storage and distribution. It has also been our prime concern that the new technologies used in all these do not degrade the environment. India has made a great strides and the green revolution has been a consequence of introduction of newer varieties of food crops, newer agriculture practices and their spread to the masses. In this article an attempt is made to foresee how a very new technology, which is being pursued vigorously all over the world, will impact the agriculture.

This Chapter introduces applications of emerging field of nanoscience and nanotechnology in agriculture.

15.2 NANOSCIENCE AND NANOTECHNOLOGY

Nanoscience is the study of phenomena and manipulation of materials at atomic, molecular and macromolecular scales, where properties differ significantly form those of the bulk materials. Nanotechnology is the design characterization, production and application of structures, devices and systems by controlling the shape and size at nanometer scale. One nanometer (nm) is equal to one-billionth of a meter. A human hair is approximately 80,000 nm wide. Atoms are below nanometer in size whereas many molecules, including some proteins, range from a nanometer upwards. Nanotechnology is an interdisciplinary area of research and development activity, which is growing at a rapid pace worldwide. It provides a stage to all researchers and technologists from chemistry, physics, material science, biology, agriculture, engineering, medicine and environmental sciences to work together. Advances in nanotechnology made it possible to build structures which are being used as sensors, filter, sunscreens, barcodes, drugs delivery systems and other devices.

15.3 BIOLOGICAL SYSTEM AND NANOTECHNOLOGY

Living organisms are built of cells that are typically 100 nm across. However, cell parts are much smaller and are in the nano size domain. The human cells (10,000 to 20,000 nanometers in diameter), organelles and large biological macromolecules such as enzymes and receptors- hemoglobin, is approximately 5 nm in diameter, while the lipid bilayer surrounding cells is on the order of 6 nm thick. Nanoscale devices and nanoscale components of larger devices are of the same size as biological entities. Nanoscale devices smaller than 50 nanometers can easily enter most cells, while those smaller than 20 nanometers can transit out of blood vessels. As a result, nanoscale devices can readily interact with biomolecules on both the cell surface and within the cell, often in ways that do not alter the behavior and biochemical properties of these molecules.

Understanding of biological process on the nanoscale level is a strong driving force behind development of nanotechnology. Size dependent physical properties of nanomaterials like optical and magnetic are the most used for biological applications. Advances in technology have been made feasible to make measurements at the sub cellular level and understanding the dynamics and mechanistic properties of molecular biomachines, allowing the direct investigation of enzyme reactions, protein dynamics, DNA, transcription and cell signaling. Nanoscale instrumentation such as AFM has allowed us to make measurement of the intermolecular mechanics of a single protein molecule, polymer molecule and small RNAs. So, Nanotechnology provides us the tools to measure and understand biosystems and biosystems offer models for inspiration for nanotechnology.

There are two basic approaches for fabricating nanomaterials "top down" and "bottom up." In top down approach one usually starts from macroscopic materials and tries to scoop out smaller devices and structures. This is the approach one use in VLSI/ULSI chips. Complementary to top down is the bottom up approach in which nanostructures are obtained by putting atoms or molecules together. Bottom up approach is used by nature since time immemorial. Nature has evolved over long times. Bio-organic molecules have complex structures with very complex dynamical behavior, typically of living cells. These cells, self assemble and form further complex structures transforming in life forms like animals and human beings. Even when damage is done to the living cells, nature has amazing ability to heal itself by self-organization e.g. when a living cell is wounded, the body reacts by sending white blood cells towards off the infections killing germs, red blood cells and protein form a seal cover over the wound. Nutrients sent to the cells, to produce new cells to replace the damaged cells. Emulating the concept and principles from the nature various techniques are applied in nanotechnology in creating new materials, devices and systems, for example, self-assembly and self-organization has inspired ideas for bottom up approach. So, biosystems offer models inspiration of nanotechnology (Figure 15.1).

15.4 NANOTECHNOLOGY: WHY SO MUCH INTEREST?

To make the material small does not mean that it has any practical use. But fact is that at the nanoscale, materials show some very interesting properties. New technologies make it possible to fabricate and control the structure of nanomaterials and possible to design materials with desired properties.

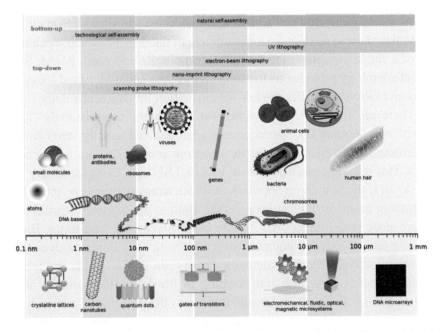

FIGURE 15.1 Size of some natural and manmade nanosized things (Source: https://upload.wikimedia.org/wikipedia/commons/2/29/Biological_and_technological_scales_compared-en.svg).

In nanomaterials, properties change nonlinearly with the size. This offers unprecedented opportunities for design of a new generation of structural and functional materials. Nanotechnology is harbinger of the next revolution with enormous opportunities. When we decrease the size of materials its physical properties change drastically as the dimensions reach 100 nm or so. The main reasons for this change in behavior are

a. An increased relative surface area leads to increased chemical activity, making nanomaterials useful as catalysts and also to use nanomaterials to improve the efficiency of fuel cells and batteries. For example, a particle of 30 nm has 5% atoms on its surface, at 10 nm 20% of its atoms and at 3 nm 50% of its atoms.

b. Confining electrons within a space of a few nanometers causes them to exhibit quantum behavior, i.e., the system would have discrete energy levels that lead to change in electrical, optical and magnetic properties. This allows us to develop products with prescribed properties. Unique electrical and optical properties together with size tunable colors enable nanostructure to be used as powerful ultra sensitive sensors identification tags.

c. Most metals are made up of crystalline grains, and the boundaries between grain, slow down or arrest the propagation of dislocations when the material is stressed, thus giving it strength. If the grains are made smaller, the grain boundary area increase and so does the strength of the material.

Thus nano devices are more efficient, fast, and light in weight, use low power consumption and may have molecular dimension.

15.5 PROGRESS IN NANOTECHNOLOGY

The progress in nanotechnology by mentioning some major advances in the field is portrayed in this section.

a. In 1959, the first reference to miniaturization was made by Richard Feyman in his famous APS lecture entitled "There is plenty of room at the bottom" at California Institute of Technology, Pasadena in which he said "*In great future we can arrange the atoms the way we wand, device will be made on atomic scale, materials will be with intelligence, motors smaller than pin head will control every thing."*

b. In 1965, Gorden Moore gave his famous "Moorie's Law": *Integrated circuit density of performance would double every 18 months*. The continuous miniaturization has brought the electronic components in the nano domain.

c. Norio Taniguchi, a Japanese researcher at the University of Tokyo, first coined the word "Nanotechnology" in 1974 at IBM in USA a technique called electron beam lithography was used to create nanostructure and devices as small as 40–70 nm in early 1970's.

d. G. Benning and H. Rohrer in 1980 at IBM Zurich Laboratory developed the scanning Tunneling Microscope and individual atoms were photographed for the first time. This powerful microscopy technique can form an image of individual atoms on a metal or semiconductor surface by scanning the tip of a needle over the surface at a height of only a few atomic diameters. G. Benning also developed the atomic force Microscope in 1984. With the invention of these imaging techniques, our understanding of the nanoworld improve dramatically, and leading to an enhance ability to control structure at the nanoscale.

e. In 1985, R. Smalley and H. Karoto discovered the fullerene molecule C_{60}.

f. In 1986, futurist K. Ethrick Drexler in his famous book "Engine of creation: The coming era of Nanotechnology" gave the idea of Molecular machine (this gives us a change to look at the nanoworld).

g. In 1993, Lijima (NEC, Japan) discovered single walled carbon nanotubes. Carbon nanotubes are only a few nanometers in diameter and can be made up to several centimeters long; giving the some unique properties they are as stiff as diamond and also conduct electricity.

15.6 APPLICATIONS OF NANOTECHNOLOGY IN AGRICULTURE

Nonmaterial and Biological system at sub cellular level are of the same size. Therefore, genetic engineers and material scientists are trying to make use of this bio conjugation where one tries to integrate nanomaterials in unprecedented ways. This prospect for nanodevices such as nanosensors, nanotubes, nanowires, nanoregisters and nanorobots for application in biological systems are being explored. Application of nanotechnology in medicine, defense, research, agriculture and food security are specially promising and there are studies to explore the possibilities of agricultural productivity enhancement. Some of the possible applications of nanotechnology are given in Figure 15.2.

15.6.1 ENHANCEMENT OF SOIL FERTILITY

Nanotechnology has found applications in enhancement of agricultural productivity using bio-conjugated nanoparticles (encapsulation) for slow release of nutrients and water, controlled release of nitrogen, characterization of

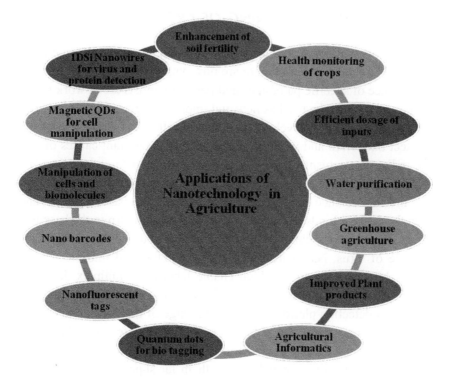

FIGURE 15.2 Different applications of nanotechnology in agriculture.

soil minerals, studies of weathering of soil minerals and soil development, micro-morphology of soils, nature of soil rhizosphere, nutrient ion transport in soil-plant system, emission of dusts and aerosols from agricultural soil and their nature, zeoponics, and precision water farming. In its stride, nano-technology converges soil mineralogy with imaging techniques, artificial intelligence, and encompass bio molecules and polymers with microscopic atoms and molecules, and macroscopic properties (thermodynamics) with microscopic properties (kinetics, wave theory, uncertainty principles, etc.), to name a few. Nanotech materials can be developed for the slow release and efficient dosage of fertilizers for plants and of nutrients and medicines for livestock. Indian chemical company Tata Chemicals is developing nan-otechnology-based crop-specific, high-value fertilizers to improve agricultural yields. The company is developing fertilizer with nanoscale particles, so that less fertilizer can be used for greater results.

15.6.2 HEALTH MONITORING OF CROPS

Nano sensors could be used to monitor crop health and when applied to the skin of a sample of plant there are now nanosensors which can signal the presence of specific pathogens and pests. Scientists are at work to see if nanoparticles can be used as new and improved pesticides, insecticides and insect repellents. Development in nanotechnology made it possible to detect bacterium and viruses, which are most important causes of disease in living organism, at early stage. The concept behind detection of a single virus is, when a virus binds to an antibody receptor on a nanode-vice, the conductance of that device will change from baseline value, and when the virus unbinds again, the conductance will return to the base line value. These discrete changes in conductance are definitive proof of the presence of virus. Nanowire sensors can be integrated into electrically addressable sensors arrays, which enable label-free multiplex detection of biological and chemical species. These nanosensors devices provide a powerful detection platform for a broad range of biological and chemical species. These nanosensors have a number of key features like direct, label free and real time electrical signal transduction, ultrahigh sensitivity and exquisite selectivity, which makes them far superior to other sensor technologies available today. These nanosensors have the unique capabilities for the detection of proteins, viruses, and DNA to the analysis of small organic molecule binding to proteins, which has the potential to impact significantly diseases diagnosis, genetic screening and drug discovery, as well as serve as powerful tools for research in many areas of agriculture and biology.

15.6.3 EFFICIENT DOSAGE OF INPUTS

Hollow nanomaterials like fullerenes (Buckyballs) and other cage like structures like dendrimers can be used for slow release and efficient dosage of fertilizers and water for plants. Nanoliter systems (lab on a chip) and other biosensors based on nanotubes, nanowires, nanomagentic particles and quantum dots offer significant advantages over conventional diagnostic methods.

15.6.4 WATER PURIFICATION AND EFFICIENT USAGE

Nano-membranes and nanoclays are portable and easily cleaned systems that purify, detoxify and desalinate water more efficiently than conventional bacterial and viral filters. Researches have developed a method of large-scale production of carbon nano-tube filters for water filtering. TiO_2 decomposes organic pollutants and my enable use of contaminated water for irrigation. Of the water used in the world, 67% is used for agriculture, 19% for industry and less than 9% for residential use.

15.6.5 GREENHOUSE AGRICULTURE

Moving agriculture into greenhouses can recover most of the water used, by dehumidifying the exhaust air and treating and re-using runoff. Additionally, greenhouse agriculture requires less labor and far less land area than open field agriculture, and provides greater independence from weather conditions including seasonal variations and drafts. Nanotechnology can help greenhouse agriculture through nanosensors for humidity, temperature and nutrients.

15.6.6   PLANT PRODUCTS

Lot of plant products can be modified by nanotechnology. Consider the potential of nanoscale innovations to affect market for rubber. Researcher in the US are designing nanoparticles to strengthen and extend the life of automobile tires as well as new material that could be used as a substitute for natural rubber. Nano-designed tires and other products require little or no rubber in the future.

15.6.7   AGRICULTURAL GEOINFORMATICS

Agriculture is inherently heterogeneous in space and time with respect to soil, crops, animals and weather. One therefore has to handle enormous amount of data, i.e., acquire it, store it, manipulate it and transfer it for optimizing both inputs and outputs in agricultural production. The future of agriculture will rely heavily on sensing, acquisition, analysis, storage and transmission of reliable and accurate data of the agricultural plants and environment for high yield and good quality. This requires successful fusion of information and communication technology. Nanotechnology through faster computers and faster communication tools like nano optoelectronic devices will make this possible.

15.6.8   QUANTUM DOT FOR BIO TAGGING APPLICATIONS

Quantum dots (QDs) have marked a great break through in many biological labeling applications with their tiny size range about 1–10 nm and outstanding fluorescent properties over current fluorophores. Taking advantage of the unique attributes of nanometer-scale semi conduct or particles, QDs provide bright, photostable, color-tunable and sharp fluorescence. Their narrow emission with the broad absorption spectra and colors controllable by varying size and composition enable multicolor labeling. Water-soluble quantum dots have been developed to improve biocompatibility allowing long-term multicolor imaging of live cells and even fluorescence correlation spectroscopy. Encapsulation within polymeric or lipid based layers and coating with a short chain of peptides also help to disguise QDs as similar sized biomolecules like proteins or nucleic acids.

15.6.9 NANO FLUORESCENT TAGS

Nanoparticles exist in the same domain as proteins, which make them suitable for bio tagging or labeling. Since nanoparticles change their optical properties like fluorescence, these characteristics may be used in bio tagging or labeling. However, size is just one of many characteristics of nanoparticles that itself is rarely sufficient if one is to use nanoparticles as biological tags. In order to interact with biological target, a biological or molecular coating or layer acting as a bioinorganic interface should be attached to the nanoparticle. Examples of biological coatings may include antibodies, biopolymers like collagen, or monolayer of small molecules that make the nanoparticles biocompatible. In addition, as optical detection techniques are wide spread in biological research, nanoparticles should either fluoresce or change their optical properties. Their approaches used in constructing nano-biomaterials are schematically presented bellow

Nanoparticles usually form the core of nano-biomaterials. It can be used as a convenient surface for molecular assembly, and may be composed of inorganic or polymeric materials. It can also be in the form of nano-vesicle surrounded by a member or layer. The shape is more often spherical but cylindrical, plate-like and other shapes are possible. The size and size distribution might be important in some cases, for example if penetration is through a pore structure of a cellular membrane is required. The size and size distribution are becoming extremely critical when quantum-sized effects are used to control material properties. A tight control of the average particle size and a narrow distribution of sizes allow creating very efficient fluorescent probes that emit narrow light in a very wide range of wavelengths. This helps with creating biomarkers with many and well distinguished colors. The core itself might have several layers and be multifunctional. For example, combining magnetic and luminescent layers one can both detect and manipulate the particles. The core particle is often protected by several monolayer of inert material, for example silica. Organic molecules that are adsorbed of chemisorbed on the surface of the particle are also used for this purpose. The same layer might act as biocompatible material. However, more often an additional layer of linker molecules is required to proceed with further fictionalization. This linear linker molecule has reactive groups at both ends. One group is aimed at attaching the linker to the nanoparticle surface and the other is used to bind various moieties like biocompatible (dextran), antibodies, fluorophores, etc., depending on the function required by the application.

15.6.10 NANO BARCODES

The ever increasing research in proteomic and genomic generate escalating number of sequence data and requires development of high through-put screening technologies. Realistically, various array technologies that are currently used in parallel analysis are likely to reach saturation when a number of array elements exceed several millions. A three-dimensional approach, based on optical "bar coding" of polymer particles in solution, is limited only by the number of unique tags one can reliably produce and detect. Single quantum dots of compound semiconductors were successfully used as a replacement of organic dyes in various bio-tagging applications. This idea has been taken one step further by combining differently sized and hence having different fluorescent colors quantum dots, and combining them in polymeric microbeads. A precise of quantum dot ratio has been achieved. The selection of nanoparticles used in those experiments had 6 different colors as well as 10 intensities. It is enough to encode over 1 million combinations. The uniformity and reproducibility of beads was high letting for bead identification accuracies of 99.99%.

15.6.11 MANIPULATION OF CELLS AND BIOMOLECULES

Nanotechnology-based platforms have been successfully used as novel detection methods for better understanding for the genetic and proteomic information on the molecular and cellular levels with high selectivity and sensitivity. Nanotechnologies for detecting biomolecules with higher sensitivity could provide an easier way to discover the unknown mechanisms in systemic biochemistry and medicinal biology. Nanotechnology-based tools also allow us to provide new methods in probing signaling pathway and spatiotemporal dynamics of cellular behavior. The single-cell detection method has been used to find the information on how a cell responds to its environmental change. Interact with other neighboring cells and express the specific genes and proteins as its response.

15.6.12 MAGNETIC QDS FOR CELLS AND BIO-MOLECULES MANIPULATION

Functionalized magnetic nanoparticles have found many applications including cell separation and probing; these and other applications are discussed in recent review. Most of the magnetic particles studied so far are spherical, which somewhat limits the possibilities to make these nanoparticles multifunctional. Alternative cylindrically shaped nanoparticles can be created by employing metal electro-deposition into nonporous alumina template. Depending on the properties of the template, nanocylinder radius can be selected in the range of 5 to 500 nm while their length can be as big as 60 μm. By sequentially depositing various thicknesses of different metals, the structures and the magnetic properties of individual cylinders can be tuned widely.

As surface chemistry for fictionalization of metal surfaces is well developed, different legends can be selectively attached to different segments. For examples, porphyrins with thiol or carboxyl linkers were simultaneously attached to the gold or nickel segments, respectively. Thus, it is possible to produce magnetic nanowires with spatially segregated fluorescent parts. In addition, because of the large aspect ratio, the residual magnetization of these nanowires can be high. Hence, weaker magnetic field can be used to drive them. It has been shown that a self-assembly of magnetic nanowires in suspension can be controlled by weak external magnetic fields. This would potentially allow controlling cell assembly in different shapes and forms. Moreover, an external magnetic field can be combined with a lithographically defined magnetic pattern ("Magnetic Trapping").

15.6.13 ONE DIMENSIONAL SI-NANOWIRES FOR VIRUS AND PROTEIN DETECTIONS

Proteins are the important part of the cell's language, machinery and structures, and understanding their functionalities is extremely important for further progress in human well being. Gold nanoparticles are widely used in immune-histochemistry to identify protein-protein-interaction. However, the multiple simultaneous detection capabilities of this technique for detection and identification are possible using single dye molecules. By combining both methods in a single nanoparticle probe one can drastically improve the multiplexing capabilities of protein probs. The group of Prof. Mirkin has designed a sophisticated multifunctional probe that is built around a 13 nm gold nanoparticle. The nanoparticles are coated with hydrophilic oligo-nucleotides containing Raman dye at one end and terminally capped with a small molecule recognition element (e.g., biotin). Moreover, this molecule is catalytically active and will be coated with silver in the solution of Ag (I) and hydroquinone. After the probe is attached to a small molecule or an antigen it is design to detect, their sub trait is exposed to silver and hydroquinone solution. A silver planting is happening close to the Raman dye, which allow for dye signature detection with a standard Raman microscope. Apart from being able to recognize small molecules this probe can be modified to contain antibodies on the surface to recognize proteins. When tested in protein array format against both small molecules are proteins, the probe has shown no-cross reactivity with the surrounding tissues. The major trend in further development of nanomaterials is to make them multifunctional and controllable by external singles or by local environment thus essentially turning them into nano-devices.

15.7 REGULATORY CONCERNS/TOXICOLOGICAL STUDIES

There are questions about the adequacy of current regulatory framework, which deals with nanotechnology. Concerns have been expressed that features like high surface reactivity and ability to cross cell membranes will adversely affect human health. Until further toxicological studies have been undertaken, human exposure to airborne nanoparticles should be avoided. If nanoparticles enter the skin, they may facilitate production of reactive molecules that could lead to cell damage. Because of more reactively and the fact that quantum effects become significant at the nanoscale, toxicity of same chemical may be different from the bulk. There are examples where nanoparticles can produce toxic effects even if the bulk substance is nonpoisonous. This is because nanoparticles have the increased surface area and enter into the human body skin inhalation and many damage the cells.

There is no information on the environmental impact of the nanoparticles. We outline a program of research required to determine the behavior of nanoparticles in the environment. In the mean time, any potential risk is managed minimizing environmental releases as far as possible. Factories and research laboratories should treat manufactured nanoparticles and nanotubes as if they are hazardous and seek to reduce or remove them from waste streams. Perhaps the greatest potential source of concentrated environmental exposure is in the near future from the application of nanoparticles to soil or water for remediation. It envisaged that nanoparticles can react with soil and ground water pollutants to leave harmful compounds. It is recommended that the use of free, manufactured nanoparticles in environmental applications, such as cleaning up contaminated land, should prohibited until appropriate research can show that potential benefits outweigh risks.

15.8   SUMMARY

Present day agriculture is facing the problem of declining trends in agricultural productivity due to various environmental threats like changes in land use particles and declining water tables, increasing temperature, excessive application of chemical and fertilizers. There are issues like sustainability of our agriculture system. It has been pointed over the nanotechnology has potential to overcome some of the problems one is facing. There is need for further investigation regarding how nanotechnology can help us in the field of agriculture. Authors are of the view that nanotechnology combined with biotechnology will bring us the next revolution in agriculture, our country should not lag behind. It is obligatory for us to train and encourage research in agricultural nanotechnology and make the society aware about the possibilities.

KEYWORDS

- **agricultural geoinformatics**
- **bio tagging**
- **biological system**
- **fullerenes**
- **magnetic QDS**
- **magnetic trapping**
- **nano-designed tires**
- **nano-membranes**
- **nanobarcodes**
- **nanoclays**
- **nanodevices**
- **Nanofertilizers**
- **nanofluorescent tags**
- **Nanomaterials**
- **nanooptoelectronic devices**
- **Nanoscience**
- **nanosensors**
- **Nanotechnology**
- **nanotoxicity**
- **quantum dots**
- **Si-nanowires**

REFERENCES

1. Alivisatos, P. (2004). The use of nanocrystals in biological detection. Nat. Biotechnol., 22, 47–52.
2. Bruchez, M., Morronne, M., Gin, P., Weiss, S., & Alivisatos, A. P. (1998). Semiconductor nanocrystals as fluorescent biological labels. Science, 281, 2013–2016.

3. Michael, F. L., De Volder, Sameh, H. T., Ray, H., Baughman, A., & John H. (2013). Carbon nanotubes: Present and future commercial applications. Science, 339(6119), 535–539.

4. Michael, J. P. (2004). Nanomaterials—the driving force, Nano Today, 12, 20.

5. Moriatry, P. (2001). Rep. Prog. Phys., 64, 297.

6. Murday, J. S. (2002). Amptiac Newsletter, 6, 5.

7. Nanoscience and nanotechnologies: opportunities and uncertainties, (2004). The Royal Society & The Royal Academy of Engineering, London.

8. Natalia, C. T., & Zhiqiang, G. (2006). Nanoparticles in biomolecular detection. Nano Today, 1, 28–37.

9. Niemeyer, C. M. (2001). Nanoparticles, proteins, and nucleic acids: biotechnology meets materials science. Angew Chem Int Ed Engl., 40, 4128–4158.

10. Potolsky, F., & Liber, C. M. (2005). Nanowire nanosensors. Materials Today, 8(4), 20.

11. Taton, T. A. (2002). Nanostructures as tailored biological probes. Trends Biotechnology, 20, 277–279.

12. Timoty, J, Raheleh, B., Daniela, R., Pocock, M., & Alexander, M. S. (2007). Biological applications of quantum dots. Biomaterials, 28, 4717–4732.

13. Wang, Y., Tang, Z., & Ktove, N. A. (2005). Bioapplication of nanosemiconductors, Nano Today, 5, 20.

14. Whitesides, G. M. (2003). The right in Nanobiotechnology. Nature Biotechnology, 21, 1161–1165.

CHAPTER 16

NANO PARTICLE BASED DELIVERY SYSTEM AND PROPOSED APPLICATIONS IN AGRICULTURE

DEEPAK KUMAR VERMA, SHIKHA SRIVASTAVA,
PREM PRAKASH SRIVASTAV, and BAVITA ASTHIR

CONTENTS

16.1 INTRODUCTION

In the recent past, various players in the field of technology have been exploring the possibility of using new science to sustain mankind, as resources become scarce and demand increases. The emergence of new nano-devices and nano-materials opens up potential novel application in agriculture and biotechnology. Use of such technology employs fundamental knowledge of biology, chemistry, biochemistry, molecular biology, chemical engineering, agronomy, etc. to decipher the science and technological knowhow to rely on nano-based technology. The term "nanotechnology" indicates a multi-disciplinary approach concerning materials, devices and systems in which at least one of three characteristic dimensions of their components is measured at the nanometric scale (nm). The nanometric scale is characterized by: (i) the atomic dimensions, (ii) the molecular dimensions, and (iii) the distance among the atoms in ordinary condensed matter.

The nano-meter sub-multiplies in the atomic world are more commonly expressed in Angstroms (A°), where $1 A° = 0.1$ nm $= 10^{-10}$ m. Lower resistance to electricity, lower melting point and faster reactions are among the basic properties to be shown by the nano-devices [24]. Reduction in size also leads to the appearance of surface effects related to the high number of surface atoms as well as to a high specific area, making these nanomaterials important from the practical point of view.

Nano-based "smart delivery systems" are most promising and recent technology that has been widely used in agriculture and medical field. Paul Ehrlich termed it as "magic bullets" loaded with specific analyte and drug to the target site, thus taking appropriate remedial action [18]. In these days, bio-degradable nanoparticles are gaining much more attention due to the site specific delivery of various biological active compounds viz. micro-macronutrients, vitamins, plant growth regulators, plant staining agents and gene [46]. Proteins, polysaccharides and synthetic bio-degradable polymers (like chitosan based nanomaterial) are bio-degradable in nature and pose an eco-friendliness to the environment. The selection of base material depends upon many factors like size, surface properties, encapsulating materials and its biocompatibility, etc. (Figure 16.1).

There are lots of emerging applications mostly in agriculture describing its role as an essential and demanding role for smart use of nano-based materials presented (Table 16.1). Torney et al., [48] reported the use of gold based nanoparticle as suitable carrier materials for recombinant DNA to target into the plant cells for plant transformation and its genetic modification, which is found to be the recent use of the nano-based composite material in the field of agriculture.

16.2   CLASSIFICATION OF NANODELIVERY SYSTEM

Nanodelivery systems (Figure 16.2) are classified into three main groups, they are; emulsions, vesicular delivery system, and lipid based nanoparticles. The two types of emulsions nanodelivery systems are

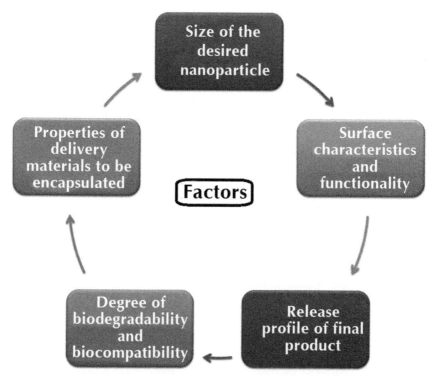

FIGURE 16.1 Factors influencing bio-degradable nanomaterials preparation.

micro-emulsions (MEs) and nano-emulsions (NEs). The three types of vesicular nanodelivery systems are liposomes, niosomes and transfersomes. The group of lipid-based nanoparticles is composed of polymeric nanoparticles and nano-porous material.

16.2.1 EMULSIONS

In the very simple form, emulsion is the mixture of two immiscible liquids, generally oil and water. Talking in nanotechnology sense these are classified into two types.

16.2.1.1 Micro-Emulsions

Micro-emulsions (MEs) are emulsion of water, oil and amphiphile (Figure 16.3) with size range in the nano-meter and with diastereomeric and

TABLE 16.1   Nanotechnology Based Advancement and Applications in Agriculture

Invention Year	Nanotechnology based Advancement	Applications in Agriculture	Ref.
2014	• Chitosan nano-formulation loaded with herbicide	• Control release of herbicide and weed management	[19]
	• Nano-based atrazine	• Weed control and management of competitive weeds	
2013	• Nano-gel formulation impregnated with phero-mone	• Management of fruit fly and other bugs	[7]
	• Silica conjugated nano-materials with naphthal-inic acid	• As a plant growth enhancer	
2012	• Nano-emulsion of neem oil	• Larvicidal agent	[34]
	• Antifungal nano-disks with amphotericin-B	• Treatment of fungal infection in wheat and chick pea	
2011	• Amino nano-composite	• Detection of organochlorine and organophosphorus in vegetables	[4]
	• Zn-Cd optical nano-sensor		
		• Detection of pesticide	
2010	• Nano-clay material	• Soil enhancer and detection water retention capacity	[43]
	• Gold nano-composite		
	• Magnetic carbon nano-tubes	• Pesticide detection in veg-etables	
	• Gold electrochemical sensors	• Smart delivery of agro-chemicals	
		• *Sclerotinia* pathogen detection	
2009	• Polyethylene glycol based nanomaterial with essential oil	• Control of pest infestation in vegetables	[25, 53]
	• Cd-Te nano-structure	• Detection of 2,4-D herbicide	
	• Polyaniline nano-based material	• Detection of parathion and chloropyrifos	
	• Nano bio-sensor	• Detection of storage infec-tion	
	• Gold nanoparticle	• Detection of pesticide	

TABLE 16.1   (Continued)

Invention Year	Nanotechnology based Advancement	Applications in Agriculture	Ref.
2008	• Nano-emulsions • Nano-starch particles	• Plant growth regulator and stress alleviator • Transporting DNA to transform plant cells	[2, 27]
2007	• Aliposome nano-sensor • Mesoporous silica nanoparticle	• Pesticide detection • Targeted delivery of DNA to plant cell for transformation	[51]
2006	• Methyl acrylic acid polymer nano-structure • TiO$_2$ Nano-structure on electrode • Porous silica nanoparticles • TiO$_2$ coated with filters	• Detection of pesticides and its analysis in vegetables • Detection of parathion residues in vegetables • Controlled delivery system for water-soluble pesticide • Photo-catalytic degradation of agrochemicals in contaminated waters	[26, 32]
2005	• Zn-Al layered nano-composite	• Control release of herbicides and pesticides	[22]
2003	• Nano-silica based soil binder	• Prevention of soil run-off • Seed blending for germination	[22]

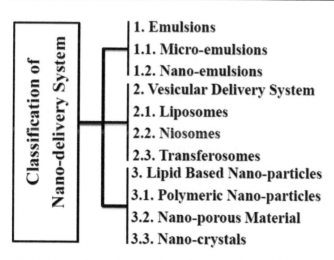

FIGURE 16.2   Flow diagram of classification of nanodelivery system.

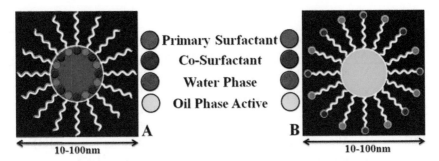

FIGURE 16.3 Structure of micro emulsion: (a) Water-oil Micro-emulsion droplet; and (b) Oil-water Micro-emulsion droplet. Source: http://www.enviroquestgpt.co.uk/content/micro-emulsion.html.

thermodynamically stable liquid containing particle with diameters of 100 nm and less [3]. These are transparent emulsions in which oil is dispersed in an aqueous medium containing surfactant with or without suitable co-surfactant. These favor formation of thermodynamically stable system with droplets of internal phase. Specific analyte/molecule carried in the micro-emulsion in the solubilized form either in the oil or the aqueous phases. MEs are being employed for the delivery of drugs as it has the ability to improve the solubility and stability of drug/chemicals [8].

16.2.1.2   Nano-Emulsions

Nano-emulsions (NEs) droplets measure between 10 and 100 nm in diameter and are typically transparent due to these sizes being on a scale smaller than the ultraviolet-visible light range. Two types of methods are commonly used in NEs synthesis, mechanical or chemical processes. Mechanical processes employ sonicators and micro-fluidizers to break larger emulsion droplets into smaller ones where as chemical method result in spontaneous formation of emulsion droplets due to hydrophobic effect of lipophilic molecules in the presence of emulsifiers [20, 31].

16.2.2 VESICULAR DELIVERY SYSTEM

16.2.2.1 Liposomes

Liposomes (Figure 16.4) are self-assembled nanoparticles formed from phospholipids in the form of a bilayer. In the lipososmes, each sub units are composed of one polar head group and other side with long chain of hydrophobic residues with microscopic size. Two monolayers fold back to back to form a liposome layer in a two dimensional manner. Bilayer formation occurs most readily when the cross-sectional areas of the head group and acyl side chain(s) are similar, as in glycerophospholipids and sphingolipids.

The hydrophobic portions in each monolayer, excluded from water, interact with each other. The hydrophilic head groups interact with water at each surface of the bilayer. Because the hydrophobic regions at its edges are transiently in contact with water, the bilayer sheet is relatively unstable and spontaneously forms a third type of aggregate: it folds back on itself to form a hollow sphere, a vesicle or liposome (Figure 16.5). These vesicles consist primarily of phospholipids (synthetic or natural), sterols and an antioxidant [10, 11, 29, 45].

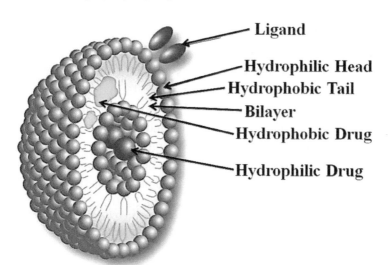

FIGURE 16.4 Structural of liposomes [6]. (Source: Bei, D., Meng, J., & Youan, B. C. (2010). Engineering nanomedicines for improved melanoma therapy: progress and promises. Nanomedicine (Lond), 5(9), 1385–1399. Used with permission of the publisher, Future Science Group.)

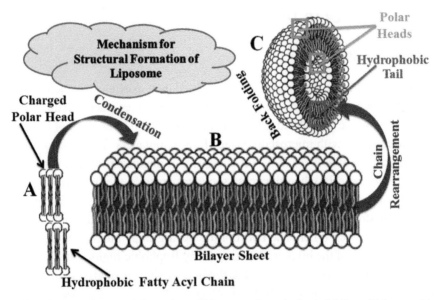

FIGURE 16.5   Structural formation of Liposome from the back folding of bilayers. (a) Basic unit of Liposome showing polar head and hydrophobic tail, (b) condensation of all basic unit to form bilayer, and (c) Back folding and chain rearrangement of bilayer to form the Liposome.

Size, number of lamellae and surface charge differentiate the liposome structures and are classified as oligo-, uni- or multi-lamellar and small, large or giant. Uni-lamellar liposomes contain a single layer and are further classified into various size ranges (Table 16.2). Multi-lamellar liposomes consist of many concentric lamellae, exhibiting an onion-like structure with diameter range between 500 nm and 5 μm [8]. Method for synthesizing liposomes includes the gentle hydration method and layer-by-layer electrostatic deposition. Liposomes have been developed for food application as a method for creating iron-enriched milk, antioxidant delivery, co-delivery of vitamin E and C with orange juice [30].

TABLE 16.2   Classification of Uni-Lamellar Liposomes [8]

Uni-lamellar liposomes	Size (in diameter)
Small	25–100 nm
Large	100–1 μm
Gaint	greater than 1 μm

16.2.2.2   Niosomes

Niosomes are the vesicles composed of nonionic surfactant (Figure 16.6). They have been mainly studied because of their advantage compared to liposomes, i.e., higher chemically stability of surfactant than phospholipid, require no condition for preparation and storage. They have no purity related problems and manufacturing costs are also low. Niosomes mostly show an ability to increase the stable entrapment of input drugs, improved bioavailability and enhanced penetration [49].

16.2.2.3   Transferosomes

Transferosomes (Figure 16.7) are the lipid vesicles containing large fractions of fatty acids. These are vesicles composed of phospholipid as their main ingredients with 10–25% surfactant and 3–10% ethanol [14]. The inventers claim that transferosomes are ultra-deformable and squeeze through pores less than 1/10th of their diameter. Higher membrane hydrophilicity and flexibility both help transferosomes to avoid aggregation and fusion, which are observed with liposomes [35].

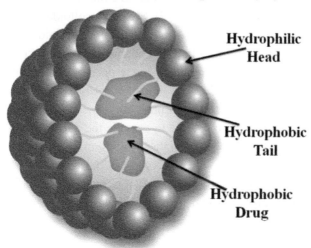

Hydrophilic Head

Hydrophobic Tail

Hydrophobic Drug

FIGURE 16.6   Structural of Niosome [6]. (Source: Bei, D., Meng, J., & Youan, B. C. (2010). Engineering nanomedicines for improved melanoma therapy: progress and promises. Nanomedicine (Lond), 5(9), 1385–1399. Used with permission of the publisher, Future Science Group.)

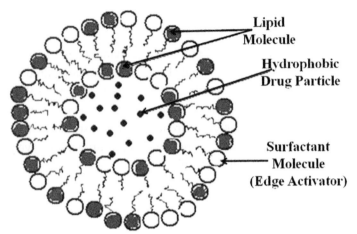

FIGURE 16.7   Structure of transferosomes [23]. (Source: Kombath, R. V., Minumula, S. K., Sockalingam, A., Subadhra, S., Parre, S., Reddy, T. R., & David, B. (2012). Critical issues related to transfersomes – novel Vesicular system. Acta Sci. Pol., Technol. Aliment., 11(1), 67–82. Used with permission from the publisher.)

16.2.3 LIPID-BASED NANOPARTICLES

Lipid based nanoparticles are mainly solid lipid nanoparticles (SLNs) and nano-structure lipid carriers (NLCs); and are similar to emulsions, with the exception that the lipids are in a solid phase [32]. SLNs are colloidal particles containing highly purified triglyceride, composed mainly of lipids that are solid at room temperature [38]. SLNs have the potential to provide controlled release of various lipophilic components due to decreased mobility of bioactive in the solid matrix [33]. NLCs are lipid nanoparticles with both crystallized and liquid phases of lipid used for nanodelivery. The inner liquid phase dissolves and entraps the bioactive to provide the advantage of chemical stability, controlled release and a higher loading efficiency [46]. There are several methods used to produce SLNs and NLCs (Table 16.3). SLNs are developed for food applications with goal of improving shelf life, the stability of quercitin, β-carotene and α-tocopherol [54].

16.2.3.1   Polymeric Nanoparticles

Polymeric nanoparticles (PNs) are made up form polymers. These are matrices of polymers and entrapped molecules surrounded by an emulsi-

TABLE 16.3 Methods Producing SLNs and NLCs [8, 33, 46]

Methods
Emulsification-sonication
Homogenization (hot and cold)
Micro-emulsion solvent emulsification-evaporation technique
Micro-fluidization
Ultra-sonication

fier or surfactant [13, 21, 37]. The material to be delivered, i.e., drugs/bioactive is manifested by various methods like in the form of dissolved substances, entrapped/encapsulated relying upon the method of preparation, nano-capsules, nanoparticles or nano-spheres can be obtained. These particles, nano-spheres and nano-capsules are designed to protect the entrapped bioactive from degradation. In the recent years, bio-degradable PNs found to the most potential drug delivery agent and as carriers of DNA in gene therapy and their ability to deliver proteins and peptides, etc. Besides this natural polymers still enjoy popularity in delivery, some of them are gums, chitosan, gelatin, sodium alginate and albumic. It is reported that protein polymer (*zein*) nanoparticles successfully entrapped essential oils with antioxidant properties [52]. Gelatin nanoparticles encapsulated polyphenol antioxidants to protect their antioxidant activity and provide controlled release [42]. PNs are engineered by emulsification followed by solvent displacement or nano-precipitation or desolvation, salting out, and emulsion evaporation methods [39]. Other methods such as pH cycling, thermal treatment, atomization spraying, and the use of supercritical fluids are mostly applied to nanoparticle synthesis for proteins and hydrocolloids.

16.2.3.2 Nano-Porous Material

A typical and relatively simple porous system is one type of dispersion classically described as material with gas-solid interfaces as the most dominant characteristic [1]. Accordingly, porous material might be classified by the size of pores or may be distinguished by different network materials. As per the classification based on the pore size, nanoparticles range from 1 nm to 1000 nm. According to IUPAC, three distinctions can be made, they are; (i) Micro-porous material (0–2 nm pores), (ii) Meso-porous material (2–50 nm pores), and (iii) Macro-porous material (>50 nm pores). Among these, the unique structural features of organically functionalized meso-porous silica nanoparticles (MSNs), such as their chemically and thermally stable structure, large surface area (>800 m^2g^{-1}), tunable pore sizes (2–10 nm) and well defined properties have made them ideal guest for hosting guest molecules of various sizes, shapes and functionalities [1].

16.2.3.3   Nano-crystals

Nano-crystals are crystals with less than 1 μm size but within the range of 10–400 nm. They are aggregates comprising several hundred to tens of thousands of atoms that combine into a "cluster." They consist of bioactive surrounded by a surfactant [16]. Nano-crystals improve the solubility of poorly water-soluble drugs/chemicals by increasing the surface area to volume ratio which increases the dissolution rate of bioactive *in vivo*. They can be made using mechanical or chemical methods [44]. Food grade starch and protein nanoparticles are being developed for various applications in food and biomedical research [12, 15, 50].

16.3   APPLICATIONS OF NANOPARTICLE-BASED DELIVERY SYSTEM IN AGRICULTURE

16.3.1   FOR EFFICIENT DELIVERY OF FERTILIZERS

Fertilizers play a pivotal role in enhancing agriculture production in India (35 to 40%). Chemicals are inorganic fertilizers formulated in appropriate concentration and combinations supply three main nutrients: nitrogen (N), phosphorus (P) and potassium (K) for various crops. However, major part of these fertilizers is lost to environment though physical, chemical processes like leaching, volatilization, emission and cannot be absorbed by plants causing economic and resource losses [41]. To enhance the nutrient use efficiency build up by nutrients in soils, nano-fertilizers are emerging as alternative approach. For controlled and sustained release of fertilizer and pesticides, nano-fertilizers are expected to be far more effective with the high surface area to volume ratio. The use of nano-fertilizers makes plant roots and microorganisms to take nutrient ions from solid phase of minerals easily and in a sustainable way. Thus it enhances soil quality by decreasing toxic effects associated with the overdose application of fertilizer [24].

Nano-fertilizers are the encapsulated fertilizers in which NPK fertilizers are entrapped in nanoparticles. Slow release of nutrients in the environment could be achieved by coating and cementing of nano and sub nano-composites [28]. For example: Nano-zeolite can act as a potential carrier for developing smart delivery fertilizers. Zeolites are group of naturally occurring minerals and has honey comb like porous layered crystal structure with the ability to exchange ions and catalyze reactions. Being made up of networks of interconnected tunnels and cages, it can be uploaded with nutrients, i.e., N, P and K, combined with other slowly dissolving ingredients containing phosphorus, calcium and a complete suite of minor and trace nutrients. Thus, nano-zeolite acts as a reservoir for nutrients that releases "on demand" [55]. It is reported that nano-composite based fertilizers consisting of N, P, K, micronutrients, mannose and amino acids has been developed that has increased the nitrogen uptake and utilization by grain crops.

Bio-degradable chitosan nanoparticles have also gain great attention as controlled release for NPK fertilizers due to their bio-absorbable and bactericidal nature [36]. A number of studies indicate that these slowly released nano-fertilizers improve grain yield and has proved to be safe for germination of cereals. Besides the delivery of primary nutrients, sulfur coated nano-fertilizers have gained much attention due to their potential role in releasing sulfur in sulfur deficient soils. Materials such as kaolin and polymeric biocompatible nanoparticles also have potential application in nano-fertilizers. When compared to bulk fertilizers, nano-fertilizers gain much more attention due to their small size, high specific surface area, reactivity, increased solubility and easy availability to plants.

Recent research reports the potential of using polymeric nanoparticles for coating of bio-fertilizer preparations for producing formulations resistant to desiccation [17]. In this way, nanotechnology has opened a new opportunity in improving the nutrient uptake efficiency and minimized costs of environmental protection.

16.3.2   AS DELIVERY OF AGROCHEMICALS

Indiscriminate use of agrochemicals viz. pesticides and insecticides since 20th century has caused environmental pollution, emergence of resistant pests and pathogens and loss of biodiversity. Encapsulation, controlled release methods and entrapment of agrochemicals has revolutionized the use of nano-composite material for agricultural use. Nano-encapsulated chemicals possesses following characteristics such as proper concentration with high solubility, stability and effectiveness, timely release in response to environmental stimuli, enhanced targeted activity and less eco-toxicity making production of crops higher and less injury to agricultural workers. Nanoparticles within the 100–250 nm size range able to dissolve in water more effectively than existing ones due to their effective surface tension quality. Use of nano-emulsions can also be better choice in terms of application of uniform suspensions of pesticides or herbicidal agents (either water or oil-based) and contain nanoparticles in the range of 200–400 nm. Formulations in the form of gels, creams, liquids, etc. have been performed to assess the effective ness of the sol-gel based nanomaterials.

16.3.3   AS NANO-HERBICIDES

Elimination of weeds can be achieved by destroying their seed banks in the soil and prevent them from germinating when weather and soil conditions become favorable for their growth is one of the most promising technology. Molecular characterization of underground plant parts for a new target domain and developing a receptor based herbicide molecule having specific binding property with nano-herbicide molecules like carbon nano-tubes capable of killing the viable and dormant underground propagates of weed seeds.

Target specific nano-encapsulated herbicide molecules are aimed for specific receptor in the roots of target weeds [9]. When these molecules enter into system by forming association with receptor, get translocate into its parts and inhibit glycolysis of food reserves in root system thus making specific weed to starve for food and get killed.

16.3.4 AS NANO-PESTICIDES

Nano-pesticides are defined as any formulation that intentionally includes elements in the nm size range and/or claims novel properties associated with the small size. The most sophisticated approach to formulate nano-scale pesticides is encapsulating the nano-scale active ingredients within "envelope" or "shell" of nano-size. The aim of nano-pesticide formulation is to protect against premature degradation and its uniform suspensions of pesticides or herbicidal agents. Polymer based nano-pesticide formulation have received the greatest attention over the last two years, followed by formulation containing inorganic nanoparticles (silica, titanium dioxide) and nano-emulsions [26].

Hydrophobic nano-silica embedded with surface materials has been successfully used for controlling insect pests and disease incidences. Properly functionalized lipophilic nano-silica gets absorbed into the cuticular lipids of insects by physio-adsorption and damages the protective wax layer and induces death by desiccation [5].

Pesticides containing nano-scale active ingredients has been introduced in market with more research on the development of new nano-pesticides formulation are coming to the world's market with innovative use of nanomaterials in the field of agriculture. Syngenta, (world's most notable biotechnology company) is using nano-emulsions in its pesticide products. It also claims its Banner MAXX fungicide which has extremely small particle size of about 100 nm. Due to small particle size, the chemical mixes completely in water. Additionally, the fungicide gets absorbed in such a way that it doesn't even get washed off by rain or irrigation. Word's 4th ranking agrochemical corporation, BASF of Germany is conducting new research on pesticide delivery to plant root and rhizosphere with new nano-based composite materials, which involves an active ingredient whose ideal particle size is between 10–150 nm. Marketed under the name Karate ZEON is a quick release microencapsulation nano-pesticide formulation containing active compound, i.e., λ-cyhalothrin which breaks open on contact with leaves. In contest, the encapsulated product "gutbuster" only bark open to release its content when it comes into contact with alternative environment such as stomach of certain insect.

16.3.5   AS DELIVERY AGENTS OF PLANT GROWTH REGULATORS (PGRS)

Plant growth regulators are natural as well as synthetic compounds capable of modifying the biochemical pathways and physiological processes in plant. Normally PGRs are synthesized throughout the plant and transported mainly by vascular system. But with any change in external environment, cell response for balancing PGRs by up or down regulation and under such circumstances, a wanted response can't be achieved as cell naturally tries to balance the change that has occurred due to external PGR application. Therefore, a target delivery system with controlled release of PGRs is essentially required for expected developmental response.

In a tissue culture experiment, addition of nano-formulation into culture medium shows a dose dependent effect on shoot length of *Asparagus officinalis*. Saponin (3.0 mg/L) and GA_3 (0.2 mg/L) formulation with variable duration of sonication (1, 3, 5, 7, 9, 11, 13 and 15 minutes) were evaluated for shoot elongation. It was observed that highest shoot lengths were found after 5 and 7 minutes sonication [40].

Saponins form self-assembled nano-sized vesicles and can act as delivery agent for different compounds. Saponin based smart delivery system consists of PGRs, i.e., auxin, GA immobilized or encapsulated with nano-structures. The surface of delivery system is functionalized with effector molecules (signal domain) in such a way that it would enable the nano-formulation to recognize the target sites in plant tissues. This approach could lead to the safe passage of PGRs under diverse environmental conditions with site targeted delivery [40].

16.3.6   FOR DELIVERING GENETIC MATERIALS INTO PLANT TISSUES

Gene transformation in plants is normally carried out by Agrobacterium mediated and micro-projectile bombardment methods. These conventional methods suffer from many drawbacks such as low transformation, high cost, low resolution and specificity [47].

Nano-biotechnology offers a new set of tools to manipulate the genes using nanoparticles, nano-fibers and nano-capsules. It is reported that when carbon nano-fibers surface modified with plasmid DNA were integrated with viable cells, successful delivery and integration of plasmid DNA took place. This integration was further confirmed from the gene expression. The process is nearly similar with microinjection method of gene delivery and hence possible with the plant cells in which the treated cells could be regenerated into whole plant that would express the introduced trait.

DNA and RNA fragments can easily be cross-linked with metallic and non-metallic nanomaterial for delivery into plant cells. Due to their nano-size, materials can easily pass through nano-scale pores of biological membrane. Ultrasonication and mild vortexing can assist the transfer of nano-fibers to carry gene into cultured plant tissues. The application of fluorescent labeled starch nanoparticles as transgenic vehicle has been reported in which the nanoparticle biomaterial was designed in such a way that it binds and transports the gene across the cell wall by inducing instantaneous pore channels in cell wall, cell membrane and nuclear membrane. It is also possible to integrate different genes on the nanoparticles at the same time. Being labeled with fluorescent marker, the imaging can be carried out with fluorescence microscope for further understanding the movement of transferred genes.

The unique feature of organically functionalized MSNs such as chemical and thermal stability, large surface area, tunnel pore size and well defined surface properties made them ideal for hosting molecules of various size, shape and functionalities. The ability of these MSNs to penetrate cell wall opens up new ways to manipulate gene expression at isolated plant cell system.

It is reported that a honeycomb like MSNs system with 3-nm pores can transport DNA and chemicals into isolated plant cells and intact leaves [48]. In MSNs system, gene and chemical inducers are loaded in the meso-pores and encapsulated with covalently bound caps. These bound caps physically block the drug from leaching out and molecules entrapped in the pores are released by introduction of uncapping triggers, such as dithio-threitol (DTT). Capping by surface-functionalized gold nanoparticles (10–15 nm size) serves as a biocompatible capping agent and prevents the gene from leaching out. The advancement in MSN system can simultaneously deliver both the gene and the chemical that trigger gene expression. It has been reported that MSN system can deliver DNA molecules carrying a marker gene and a chemical that is needed for transgene expression into plant simultaneously and release the encapsulated chemical in a controlled manner to trigger the expression of co-delivered transgene into the cell.

Further developments such as pore enlargement and multi function-alization of MSN systems may offer new possibilities in target-specific delivery of proteins, nucleotides and chemicals in plant biotechnology.

16.4 CONCLUSIONS, FUTURE AND OPPORTUNITIES

Despite contribution of highest gross domestic product to the world, agriculture is still in susceptible condition of meeting the demand of food supply of ever increasing population. There is also an urge need of environmental sustainability with food security without harming the soil health, quality and ecosystem. Anthropogenic as well as human and industrial activities have threatened the natural environment with huge input of toxic chemicals, xenobiotic and other non-biodegradable prod-ucts. Also agricultural production is continuously being challenged by both biotic and abiotic losses which counts to be 40% of the total loss, still the food supply to the demand is at easy pace. Hence, use of excess agrochemicals for increasing the productivity is the most and that leads to deterioration of soil quality, flora and fauna as well as incidence of higher herbicide and pesticide resistance in weeds and insects-pathogens. So, there is an alarming need of ecosystem homeostasis and natural restora-tion through *"environmental 3 rule"* which includes *Reduce, Reuse and*

Recycle, where deployment of eco-friendly technology and systems is the most required. In this context, use of nanotechnology has been emerged as a basic tool for technological advancement mitigating the environmental claims, management of pest-pathogen stresses, detection of diseases, nutrient status in agricultural crops, enhancing crop yields, nano-based delivery systems for increased fertilization, water soluble pesticide, plant growth regulators, etc.

Use of nano-silica component for preventing soil run-off, Zn-Al nano-materials for controlled release of herbicides, Ti-based nano-tube for detection of diseases in vegetables and photo-activation based biodegradation of xenobiotics, nano-emulsion for plant growth promotion and stress alleviation, poly-ethoxy glycol (PEG) based nanomaterial embedded with garlic oil for control of pest incidence, gold nanoparticle for pesticide detection, carbon nano-tube for smart delivery of agro-chemicals are some of the most promising achievements incurred in the field of nanotechnology. Recently the essence of nanotechnology has been tested in biotechnology involving use of nano-gold/titanium and most interestingly chitosan based nanodelivery system for plant transformation and genetic engineering for crop biotechnology has been a subject of immense interest among agricultural researcher that have completely replaced the traditional delivery system of genetic material to plant tissues.

Use of nanotechnology based electronic devices like sensors are also being used for the purpose of environmental monitoring for detection of contaminants and with due advancements, nanotechnology has become an integrated part of agriculture. With the list of nano-based products as a tool for agricultural advancement, the mere on site (*in-situ*) application is still in juvenile stage. Exploration of nanomaterials for delivery of metabolite of interest throughout the plant vascular system with target delivery approach is still a question and a possible solution can bring a revolution in the alleviation of stress in crops and their physiological disorders, ultimately helping to make agriculture sustainable. There is an ongoing research effort in formulating low cost, efficient, low dose, selective nanomaterials and it's scaling up process for its development as well assure in agricultural production. Apart from its use, there is still a better understanding of its post use effect is required and known to be a major hurdle in the research of nanotechnology. With eyeing towards sustainable development, uses of nano-based materials have a promising future as well as great value for making sustainable agriculture and eco-friendly environment.

KEYWORDS

- agriculture
- bio-degradable
- bioactive
- biotechnology
- emulsions
- encapsulation
- liposomes
- nano-capsule
- nanodelivery
- nano-fertilizer
- nano-fiber
- nanomaterials
- nanoparticles
- nano-pesticide
- nano-scale
- nano-size
- nanotechnology
- polymer

REFERENCES

1. Afzali, A., & Maghsoodlou, S. (2015). Engineered nanoporous materials: a comprehensive review. In: Physical Chemistry Research for Engineering and Applied Sciences by Pearce, E. M., Howell, B. A. Pethrick, R. A., & Zaikov, G. E. (Eds.). Volume 1, Principles and Technological Implications, Apple Academic Press, pp. 318–350.

2. Alemdar, A., & Sain, M. (2008). Isolation and characterization of nanofibers from agricultural residues-Wheat straw and soy hulls. Bioresour. Technol., 99, 1664–1671.
3. Amnon, C., Sintov, & Haim, V. (2007). A microemultion based delivery system for the dermal delivery of therapeutics. Inno. Pharmace. Tech., 23, 68–72.
4. Bakar, N. A., Salleh, M. M., Umar, A. A., & Yahaya, M. (2011). The detection of pesticides in water using ZnCdSe quantum dot films. Adv. Nat. Sci: Nanosci. Nanotech., 2.
5. Barik, T. K., Sahu, B., & Swain, V. (2008). Nanosilica-from medicine to pest control. Parasitol. Res., 103, 253–258.
6. Bei, D., Meng, J., & Youan, B. C. (2010). Engineering nanomedicines for improved melanoma therapy: progress and promises. Nanomedicine (Lond), 5(9), 1385–1399.
7. Bhagat, D., Samanta, S. K., & Bhattacharya, S. (2013). Efficient management of fruit pests by pheromone nanogels. Sci. Rep., 3, 1294.
8. Bonifácio, B. V, Silva, P. V., Ramos, M. A. S., Negri, K. M. S., Bauab, T. M., & Chorilli, M. (2014). Nanotechnology-based drug delivery systems and herbal medicines: a review. Inte. J. Nanomed., 9, 1–15.
9. Chinnamuthu, C. R., & Kokiladevi, E. (2007). Weed management through nanoherbicides. In: Application of Nanotechnology in Agriculture by Chinnamuthu, C. R., Chandrasekaran, B., & Ramasamy, C. (Eds.), Tamil Nadu Agricultural University, Coimbatore, India.
10. Chorilli, M., Leonardi, G. R., Oliveira, A. G., & Scarpa, M. V. (2004). Lipossomas em formulações dermocosméticas [Dermocosmetic liposome formulations]. Infarma., 16(7–8), 75–79.
11. Chorilli, M., Rimério, T. C., Oliveira, A. G., & Scarpa, M. V. (2007). Estudo da estabilidade de lipossomas unilamelares pequenos contendo cafeína por turbidimetria [Study of the stability of small unilamellar liposomes containing caffeine turbidimetric]. Rev Bras Farm., 88(4), 194–199.
12. de Mesquita, J. P., Donnici, C. L., Teixeira, I. F., & Pereira, F. V. (2012). Bio-based nanocomposites obtained through covalent linkage between chitosan and cellulose nanocrystals. Carbohydr. Polym., 90(1), 210–217.
13. des Rieux, A., Fievez, V., Garinot, M., Schneider, Y. J., & Preat, V. (2006). Nanoparticles as potential oral delivery systems of proteins and vaccines: a mechanistic approach. J. Control. Release., 116(1), 1–27.
14. Dubey, V., Mishra, D., & Jain, A. A. (2006). Transdermal Delivery of a Pineal Hormone: Melatonin Via Elastic Liposomes. Biomater., 27, 3491–3496.
15. Flauzino Neto, W. P., Silverio, H. A., Dantas, N. O., & Pasquino, D. (2013). Extraction and characterization of cellulose nanocrystals from agro-industrial residue soy hulls. Ind Crops Prod., 42, 480–88.
16. Florence, A. T. (2005). Nanoparticle uptake by the oral route: Fulfilling its potential? Drug Discov. Today: Technol 2(1): 75–81.
17. Ghormade, V., Deshpande, M. V., & Paknikar, K. M. (2011). Perspectives for nanobiotechnology enabled protection and nutrition of plants. Biotechnol Adv., 29, 792–803.
18. Gonzalez-Melendi, P., Fernandez Pacheko, R., Coronado, M. J., Corredor, E., Testilano, P. S., Risueno, M. C., Marquina, C., Ibarra, M. R., Rubiales, D., & Perez-De-Luque, A. (2008). Nanoparticles as Smart Treatment-delivery Systems in Plants:

Assessment of Different Techniques of Microscopy for their Visualization in Plant Tissues. Ann. Bot., 101, 187–195.

19. Grilloa, R., Pereira, A. E. S., Nishisaka, C. S., de, L. R., Oehlke, K., Greiner, R., & Fraceto, L. F. J. (2014). Chitosan nanoparticle based delivery systems for sustainable agriculture. Hazard. Mater., 278, 163–171.

20. Guzey, D., & McClements, D. J. (2006). Formation, stability and properties of multilayer emulsions for application in the food industry. Adv. Colloid Interface Sci., 128–30, 227–248.

21. Hunter, A. C, Elsom, J., Wibroe, P. P., & Moghimi, S. M. (2012). Polymeric particulate technologies for oral drug delivery and targeting: a pathophysiological perspective. Nanomed., 8, 5–20.

22. Hussein, M. Z., Yahaya, A. H., Zainal, Z., & Kian, L. H. (2005). Nanocomposite-based controlled release formulation of an herbicide, 2,4-dichlorophenoxyacetate in capsulated in zinc-aluminum-layered double hydroxide. Sci. Technol. Adv. Mater, 6, 956–962.

23. Kombath, R. V., Minumula, S. K., Sockalingam, A., Subadhra, S., Parre, S., Reddy, T. R., & David, B. (2012). Critical issues related to transfersomes – novel Vesicular system. Acta Sci. Pol., Technol. Aliment., 11(1), 67–82.

24. Kundu, S., Huitink, D., & Liang, H. (2010). Formation and catalytic application of electrically-conductive Pt nanowires. J Phys Chem., 114(17), 7700–7709.

25. Lisa, M., Chouhan, R. S., Vinayaka, A. C., Manonmani, H. K., & Thakur, M. S. (2009). Gold nanoparticles based dipstick immunoassay for the rapid detection of dichlorodiphenyltrichloroethane: an organochlorine pesticide. Biosens. Bioelectron., 25, 224–227.

26. Liu, F., Wen, L. X., Li, Z. Z., Yu, W., Sun, H. Y., & Chen, J. F. (2006b). Porous hollow silica nanoparticles as controlled delivery system for water soluble pesticide. Mat. Res. Bull., 41, 2268–75.

27. Liu, J., Wang, F., Wang, L., Xiao, S., Tong, C., Tang, D., Liu, X., & Cent, J. (2008). Preparation of fluorescence starch-nanoparticle and its application as plant transgenic vehicle. South Univ. Technol., 15, 768–773.

28. Liu, X., Feng, Z., Zhang, S., Zhang, J., Xia, Q., & Wang, Y. (2006a). Preparation and testing of cementing nano-subnano composites of slow or controlled release of fertilizers. Sci. Agr. Sin., 39, 1598–1604.

29. Madrigal-Carballo, S., Lim, S., Rodriguez, G., Vila, A. O., & Krueger, C. G. (2010). Biopolymer coating of soybean lecithin liposomes via layer-by-layer self-assembly as novel delivery system for ellagic acid. J. Funct. Foods, 2(2), 99–106.

30. Marsanasco, M., Marquez, A. L., Wagner, J. L., Alonso, S., & Chiaromoni, N. S. (2011). Liposomes as vehicles for vitamins E and C: an alternative to fortify orange juice and offer vitamin C protection after heat treatment. Food Res. Int., 44(9), 3039–3046.

31. McClements, D. J., & Li, Y. (2010). Structured emulsion-based delivery systems: controlling the digestion and release of lipophilic food components. Adv. Colloid Interface Sci., 159(2), 213–228.

32. McMurray, T. A., Dunlop, P. S. M., & Byrne, J. A. (2006). The photocatalytic degradation of atrazine on nanoparticulate TiO2 films. J. Photochem. Photobiol A: Chem., 182, 43–51.

33. Mehnert, W., & Mader, K. (2012). Solid lipid nanoparticles. Adv. Drug Deliv. Rev., 64, 83–101.

34. Milani, N., McLaughlin, M. J., Stacey, S. P., Kirby, J. K., Hettiarachchi, G. M., Beak, D. G., & Cornelis, G. (2012). Fate of Zinc Oxide Nanoparticles Coated onto Macronutrient Fertilizers in an Alkaline Calcareous Soil. J. Agric. Food Chem., 25, 3991–3998.

35. Mozafari, M. R., & Khosravi-Darani K. (2007). An Overview of Liposome Derived Nanocarrier Technologies. In: Nanomaterials and Nanosystems for Biomedical Applications by Mozafari, M. R. (Ed.). Springer. pp. 113–123.

36. No, H. K. (2007). Applications of chitosan for improvement of quality and shelf life of foods: A review. J. Food Sci., 72(5), 87–100.

37. Plapied, L., Duhem, N., des Rieux, A., & Preat, V. (2011). Fate of polymeric nanocarriers for oral drug delivery. Curr. Opin. Colloid Interface Sci., 16(3), 228–237

38. Puri, A., Loomis, K., Smith, B., Lee, J. H., Yavlovich, A., Heldman, E., & Blumenthal, R. (2009). Lipid-based nanoparticles as pharmaceutical drug carriers: from concepts to clinic. Crit. Rev. Ther. Drug Carrier Syst., 26(6), 523–580.

39. Sabliov, C. M., & Astete, C. E. (2008). Encapsulation and controlled release of antioxidants and vitamins. In: Delivery and Controlled Release of Bioactives in Foods and Nutraceuticals by Garti, N (Ed.). Cambridge: Woodhead Publ., pp. 297–330.

40. Saharan, V. (2010). Effect of gibberellic acid combined with saponin on shoot elongation of Asparagus officinalis. Biologia. Plant, 54(4), 740–742.

41. Saigusa, M. (2000). Broadcast application versus band application of polyolefin-coated fertilizer on green peppers grown on andisol. J. Plant Nutr., 23, 1485–1493.

42. Shutava, T. G., Balkundi, S. S., & Lvov, Y. M. (2009). (−)-Epigallocatechin gallate/gelatin layer-by-layer assembled films and microcapsules. J. Colloid Interface Sci., 330(2), 276–283.

43. Singh, S., Singh, M., Agrawal, V. V., & Kumar, A. (2010). An attempt to develop surface plasmon resonance based immunosensor for Karnal bunt (Tilletia indica) diagnosis based on the experience of nano-gold based lateral flow immuno-dipstick test. Thin Solid Films., 519, 1156–1159.

44. Sun, B., & Yeo, Y. (2012). Nanocrystals for the parenteral delivery of poorly water-soluble drugs. Curr. Opin. Solid State Mater. Sci., 16(6), 295–301

45. Tamjidi, F., Shahedi, M., Varshosaz, J., & Nasirpour, A. (2013). Nanostructured lipid carriers (NLC): a potential delivery system for bioactive food molecules. Inno. Food Sci. Emerg. Technol., 19, 29–43.

46. Tarafdar, J. C., Sharma, S., & Raliya, R. (2013). Nanotechnology: Interdisciplinary science of applications. Afr. J. Biotech., 12(3), 219–226.

47. Taylor, N. J., & Fauquet, C. M. (2002). Microparticle bombardment as a tool in plant science and agricultural biotechnology. DNA Cell Biol., 21, 963–977.

48. Torney, F., Trewyn, B. G., Lin, V. S., & Wang, K. (2007). Mesoporous silica nanoparticles deliver DNA and chemicals into plants. Nat. Nanotechnol., 2, 295–300.

49. Tripathi, P. K., Choudary, S. K., Srivastva, A., Singh, D. P., & Chandra, V. (2012). Niosomes: an study on Noval drug delivery system-A review. Inter. J. Pharmac. Res. Develop., 3, 100–106.

50. Tzoumaki, M. V., Moschakis, T., & Biliaderas, C. G. (2011). Mixed aqueous chitin nanocrystal–whey protein dispersions: microstructure and rheological behavior. Food Hydrocoll., 25(5), 935–942.

51. Vamvakaki, V., & Chaniotakis, N. A. (2007). Pesticide detection with a liposome-based nano-biosensor. Biosens. Bioelectron., 22, 2848–2853.
52. Wu, Y., Luo, Y., & Wang, Q. (2012). Antioxidant and antimicrobial properties of essential oils encapsulated in zein nanoparticles prepared by liquid–liquid dispersion method. Food Sci. Technol., 48(2), 283–290.
53. Yao, K. S., Li, S. J., Tzeng, K. C., Cheng, T. C., Chang, C. Y., Chiu, C. Y., Liao, C. Y., Hsu, J. J., & Lin, Z. P. (2009). Fluorescence Silica Nanoprobe as a Biomarker for Rapid Detection of Plant pathogens. Adv. Mater. Res., 79, 513–516.
54. Zambrano-Zaragoza, M. L., Mercado-Silva, E., Ramirez-Zamorano, P., Cornejo-Villegas, M. A., Gutierrez-Cortez, E., & Quintanar-Guerrero, D. (2013). Use of solid lipid nanoparticles (SLNs) in edible coatings to increase guava (Psidium guajava L.) shelf-life. Food Res. Int., 51(2), 946–953.
55. Zhang, F., Wang, R., Xiao, Q., Wang, Y., & Zhang, J. (2006). Effects of slow/con-trolled-release fertilizer cemented and coated by nanomaterials on biology. II. Effects of slow/controlled-release fertilizer cemented and coated by nanomaterials on plants. Nanosci., 11, 18–26.

CHAPTER 17

NANOTECHNOLOGY: POLLUTION SENSING AND REMEDIATION

VEDPRIYA ARYA

CONTENTS

17.1   INTRODUCTION

Environmental nanotechnology plays a key role in shaping of current environmental engineering and science. The nanoscale has stimulated the development and use of novel and cost-effective technologies for remediation, pollution detection, catalysis and others. However, there is also a wide debate about the safety of nanoparticles and their potential impact on environment and biota. Especially the new field of nano-toxicology has received a lot of attention in recent years. There is the huge hope that nano-technological applications and products will lead to a cleaner and healthier environment. Maintaining and re-improving the quality of water, air and soil, so that the Earth will be able to support human and other life sustainably, are one of the great challenges of our time.

The scarcity of water, in terms of both quantity and quality, poses a significant threat to the well-being of people, especially in developing countries. Great hope is placed on the role that nanotechnology can play in providing clean water to these countries in an efficient and cheap way. On the other hand, the discussion about the potential adverse effects of nanoparticles has increased steadily in recent years and is a top priority in agencies all over the world. This study was intended to give an overview of the various aspects of nanotechnology and the environment, mainly looking at it from the side of applications rather than from the risk side [3, 4, 8].

Nanotechnology has potential to significantly affect the environment through the understanding and control of the emissions from a wide range of sources. A new and more advanced term is sometimes used '*Green nanotechnology*' that reveals an idea of minimizing the production of the undesirable byproducts and remediation of existing waste sites and polluted water sources. Nanotechnology has the potential to remove the finest contaminants from water supplies and air as well as continuous measure and mitigate pollutants in the environment [8].

17.2   NANOTECHNOLOGY APPLICATIONS IN AIR POLLUTION

The pollution is a major challenging problem in front of the world. Various strategies and plans are currently working on this problem but nanotechnology is gaining a special impact in the pollution detection and management.

17.2.1   AIR POLLUTION DETECTION

17.2.1.1   By Nanoparticles

Mesoporous (MnO_2) materials with gold nanoparticles are able to absorbs and decompose the volatile organic compounds (VOC's), which are the environmental pollutants. These nanoparticles have high surface area so more adsorption of the volatile organic compounds takes place. Moreover, a large number of oxygen atoms are contained in the MnO_2 lattices can takes part in the breakdown of pollutants. These structures eliminate the acetaldehyde in an hour at room temperature, producing CO_2 as a byproduct [3].

17.2.1.2 Solid State Gas Sensors (SSI) for Air Pollution Monitoring

These sensors are light weight, extremely small in size so can fit anywhere and can receive data from any location. They are also more rugged, robust, and low cost alternatives. Solid state gas sensors (SSI) are the cost effective air pollution monitoring system. Air pollution monitoring is a technique composed of the solid state gas sensors (SSI) linked via. global positioning systems (GPS) [3].

17.2.2 AIR POLLUTION MANAGEMENT

There are two major ways in which nanotechnology is being used to reduce air pollution:
- **Catalysts** are currently in use and constantly being improved upon
- **Nano-structured membranes** are under development.

Catalysts can be used to enable a chemical reaction (which changes one type of molecule to another) at lower temperatures or make the reaction more effective. Nanotechnology can improve the performance and cost of catalysts used to transform vapors escaping from cars or industrial plants into harmless gasses. That's because catalysts made from nanoparticles have a greater surface area to interact with the reacting chemicals than catalysts made from larger particles. The larger surface area allows more chemicals to interact with the catalyst simultaneously, which makes the catalyst more effective.

Nanostructured membranes, on the other hand, are being developed to separate carbon dioxide from industrial plant exhaust streams. The plan is to create a method that can be implemented in any power plant without expensive retrofitting [8].

17.2.2.1 Gold Nanoparticles in Manganese Oxide Cleans VOCs from Air

In addition to nitrogen oxides and sulfur oxides, many volatile organic compounds (VOCs) in air contribute to smog and high ozone levels, as well as potentially damaging human health. Clean-air laws are thus rightly continuing to become stricter. Most modern air-purification systems are based on photocatalysts, adsorbents such as activated charcoal, or ozonolysis. However, these classic systems are not particularly good at breaking down organic pollutants at room temperature. Japanese researchers have now developed a new material that very effectively removes VOCs as well as nitrogen and sulfur oxides from air at room temperature. As reported in the *Journal Angewandte Chemie*, their system involves a highly porous manganese oxide with gold nanoparticles grown into it.

One secret to the success of this new material is the extremely large inner surface area of the porous manganese oxide, which is higher than all previously known manganese oxide compounds. This large surface area offers the volatile molecules a large number of adsorption sites. Moreover, the adsorbed pollutants are very effectively broken down. There is clearly plenty of oxygen available for oxidation processes within the manganese oxide lattice. Degradation on the surface is highly effective because free radicals are present there. Presumably, oxygen from air dissociates on the gold surface to replace the consumed oxygen atoms in the lattice structure. This process only works if the material is produced in a very specific manner: The gold must be deposited onto the manganese oxide by means of vacuum-UV laser ablation. In this technique, a gold surface is irradiated with a special laser, which dislodges gold particles through evaporation. These gold particles have unusually high energy, which allows them to drive relatively deep into the surface of the manganese oxide. This process is the only way to induce sufficiently strong interactions between the little clumps of gold and the manganese oxide support [8].

17.3 NANOTECHNOLOGY APPLICATIONS IN WATER POLLUTION

Nanotechnology is being used to develop solution to three different problems in water quality. One challenge is the removal of industrial wastes such as cleaning solvent called TCE, from the ground water. Nanoparticles can be used to convert the contaminating chemicals through a chemical reaction to make it harmless. A new development in controlling the water pollution is the manufacturing of the water filters that are only few nanometer in diameter.

Conventional water filters cannot remove virus cells from the impure water but that nanosized water filter is capable of removal of virus like impurities from the water. The major applications of nanotechnology to remove water pollution are:

- Using iron nanoparticles to clean up carbon tetrachloride pollution in ground water.
- Filters capable of removing viruses.
- Nanoparticles that can absorb radioactive particles polluting ground-water.
- Coating iron nanoparticles allow them to neutralize dense, hydrophobic solvents polluting ground-water.
- Using nanowire mats to absorb oil spills.
- Using iron oxide nanoparticles to clean arsenic from water wells.
- Using gold tipped carbon nanotubes to trap oil drops polluting water.
- Deionization method uses electrodes composed of nano-sized fibers to remove salt and metals in water.

17.3.1 FE-NANOPARTICLES FOR CLEANING OF WATER

The use of nanosized particles of iron for cleaning up contaminants in ground water, soil and sediments is one of the most recent techniques. However, there are some unanswered questions about the appropriate and optimal implementation of the nano-iron. Moreover, there are some doubts regarding the stability of these nanoparticles.

Tratnyek and Johnson [7] observed the behavior of iron nanoparticles ranging from 10–100 nm and measured how fast these nanoparticles degrade the CCl_4, which is a chemical used in manufacturing of cleaning fluids and degreasing agents. In a few locations, spills of these liquids are present in the soil and contaminate a large area of contaminated ground water and soil. Iron oxide with a magnetite shell high in sulfur, quickly and effectively degraded CCl_4 to a mixture of relatively harmless products. It is possible to emplace nanosized iron deep into the subsurface by injecting it through deep wells. This approach may be suitable for remediation of very deep plumes of CCl_4 contaminated ground water [8].

17.3.2 NANOFILTRATION AND DESALINATION

Nanofiltration membrane technology is already widely applied for removal of dissolved salts from salty water, removal of micro pollutants, water softening, and wastewater treatment. Nanofiltration membranes selectively reject substances, which enable the removal of harmful pollutants and retention of nutrients present in water that are required for the normal functioning of the body. It is expected that nanotechnology will contribute to improvements in membrane technology. Source materials for nanofilters include naturally occurring zeolites and attapulgite clays, which can now be manipulated on the nanoscale to allow for greater control over pore size of filter membranes [2, 4].

17.3.3 NANOCATALYSTS AND MAGNETIC NANOPARTICLES

Desalination is a process of removal of excess salts from the sea water or other polluted water. The nanocatalysts and magnetic nanoparticles are used for heavily polluted water to make it suitable for drinking, sanitation, and irrigation. The catalytic particles can chemically degrade the pollutants whereas the magnetic nanoparticles, when coated with different compounds that have a selective affinity for diverse contaminating substances, could be used to remove pollutants, including arsenic, from water. The catalysts such as titanium dioxide (TiO_2) [5] and magnetic iron nanoparticles can be used to degrade organic pollutants and remove salts and heavy metals from liquids for example a nanoscale zero-valent iron as a catalyst is used to remove arsenic from groundwater.

Recently, an iron-storage protein that consists of a native nano-size iron oxide core is developed that may serve as catalysts in chemical degradation processes of common contaminants. Brazilian researchers have developed super paramagnetic nanoparticles that, coated with polymers, can be spread over a wide area in dust like form; these modified nanomagnets would readily bind to the pollutant and could then be recovered with a magnetic pump. Because of the size of the nanoparticles and their high affinity for the contaminating agents, almost 100% of the pollutant would be removed. The magnetic nanoparticles and the polluting agents would be separated, allowing for the reuse of the magnetic nanoparticles and for the recycling of the pollutants [3].

17.3.4 NANOSENSORS FOR WATER POLLUTION DETECTION

The new sensor technologies combine micro- and nanofabrication technology to create small, portable, and highly accurate sensors to detect chemical and biochemical parameters in water. Nanosensors are made to detect the pollutants in the water samples. A new sensor technology called as *'Smart sensors'* are being developed that are used for water pollution [6].

This smart sensor consists of ISFET sensors (H^+ ion sensitive field effect transistors), an operating circuit and analog digital converter. The analysis is performed by taking into account the presence in the measured solution of other than H^+ ions so called 'disturbing ions' (Figure 17.1). A silicon microsystem that consists of data processing circuits is also present. This data processing unit is responsible for the speed and quality of analog data conversion from chemical sensors. It uses the neutral network for the identification and measurement of the concentration of disturbing ions. Here, the disturbing ion is the contamination and ion other than water ions. Another important sensor is the *'mercury nanosensor'* that uses a colorimetric method for the identification and characterization of the mercury nanoparticles. This provides a simple and sensitive way to detect the mercury in water samples. Colorimetric methods are convenient because analytes can be detected with the naked eyes by observing the color changes in the reagent solutions. This method can detect the mercury ions at the concentration lower than 100 nm in the aqueous samples. This method is based on the DNA-functionalized gold nanoparticles so this procedure is highly selective. In this method, the gold nanoparticles are functionalized with one of the two types of thiolated DNA sequences. These sequences are complementary except for a single thymidine-thymidine mismatch [3].

In the absence of the mercury ions, DNA-functionalized nanoparticles aggregate to form duplex and the mismatch get dissociated at 46°C. This melting give rise to a dramatic purple to red color change. But in the presence of the mercury ions, strengthening of the complex takes place and raises their melting temperature. These effects are mediated by the mercury ions coordinating with the pairs of the mismatched thymidine

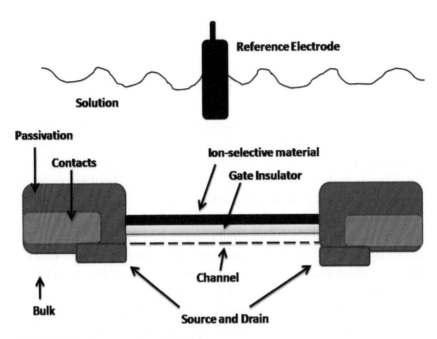

FIGURE 17.1   Cross-section of ISFET sensor.

groups. This method is only sensitive for the mercury impurity in the water samples [8].

17.3.5 CARBON NANOTUBES FOR THE DETECTION OF IMPURITIES IN RIVER WATER

The high hydrophobicity and tendency of the nanotubes to aggregate helps in the detection of the contaminants in the water. Natural organic matter (NOM) present in the river water could aid the dispersion of carbon nanotubes [1]. Carbon nanotubes are stabilized in a aqueous phase by well characterized surfactants and polymers. In the solution containing the pure water, the multiwalled carbon nanotubes can settled quickly and water became transparent in under an hour. All remaining (impure) solutions retained some coloration, indicating that some nanotubes remain in suspension.

TEM confirms that most MWNT's in NOM containing samples are not clustered but are present individually.

17.4 CONCLUSIONS

This Chapter presents was an overview of the various aspects of nanotechnology and the environment, mainly looking at it from the side of applications rather than from the risk side. It should have become clear that nanotechnology in general and nanoparticles in particular will have important impacts on various fields of environmental technology and engineering. However, we should always keep in mind that nanotechnology has a Janus face and that each positive and desired property of nanomaterials could be problematic under certain conditions and pose a risk to the environment. A careful weighing up of the opportunities and risks of nanotechnology with respect to their effects on the environment is therefore needed.

17.5 SUMMARY

Environmental nanotechnology is an emerging area of research in concern with environmental pollution. This is such a smart technology by which we can find the eco-friendly solutions of environmental problems. Nanosensing, desalination, nanofiltration, smart nanosensors are some of the smart techniques that help to eradicate environmental hazards. However, risk factors are also associated with these technologies, but it can be minimized by using cheap and eco-friendly approach. The concepts in this Chapter deal with the environmental perspective of an emerging area of science, i.e., nanotechnology.

KEYWORDS

- air pollution
- desalination
- environment nanotechnology
- magnetic nanoparticles

- **nanofiltration**
- **nanosensors**
- **nanotechnology**
- **nanotubes**
- **water pollution**

REFERENCES

1. Gould, P. (2007). Trapped dyes alter nanotube optics: optical properties. Nanotoday. 2, 3–13.
2. Hilal, N., Al-Zoubi H., Darwish N. A., Mohammad A. W., & Abu Arabi, M. (2004). Desalination, 170, 281.
3. Novak, B. (2008). Nanotechnology – Environmental Aspects. Wiley-VCH-Verlag GmbH and Company, Weinheim. Vol. 2, 1–16.
4. Srivastava, O., Srivastava, N., Talapatra, S., Vajtai, R., & Ajayan, P. M. (2004). Nat. Matter, 3, 610.
5. Strini, A., Cassese, S., & Schiavi, L. (2005). Applied Catalysis, B61, 90–97.
6. Szermer, M., Daniel, M., & Napieralski, A. (2003). Design and modeling of smart sensor dedicated for water pollution monitoring. In: NANOTECH Conference, San Francisco, CA, USA, Vol. 1, 110–114.
7. Tratnyek, P. G., & Johnson, R. L. (2006). Nanotoday, 1, 1–13.
8. Yunus, I. S., Harwin, Adi, K., Dendy, A., & Antonius, I. (2012). Nanotechnologies in water and air pollution treatment. Environmental Technology Reviews, 1(1), 136–148.

CHAPTER 18

NATURAL PIGMENTS AND DYE-SENSITIZED SOLAR CELLS (DSSC): A CORRELATION

BRAJESH KUMAR

CONTENTS

18.1   INTRODUCTION

At present, solar energy is considered the most promising renewable power source due to its easily exploitable, inexhaustible, quiet, and adjustable to enormous applications. Every day, the Sun shines on the Earth and providing around 3×10^{24} J of green energy per year, which exceeds by a factor of 10^4 the present global population consumption. A simple calculation leads self-evidently to the conclusion that covering only around 0.1% of the Earth's surface by means of energy conversion devices having an efficiency of about 10% would satisfy the present global energy needs. These encouraging numbers are inducing the scientific community to make even greater efforts towards the direction of improving solar energy conversion technologies as well as proposing new, intriguing solutions. The sun can provide the ultimate solution to the challenge of a sustainable energy supply. In one hour, the earth receives 13.6 TW from the sun, more than what we currently consume in a whole year (13 TW) [1]. The conversion of solar radiation to electrical energy (photovoltaics) has become more and more important, because sunlight is a clean and limitless energy source compared to the traditional fossil energy sources.

Recently, dye-sensitized solar cells (DSSC) are third generation devices, emerging as one of the most promising low cost photovoltaic technologies, addressing clean, secure, and efficient solar energy conversion to electricity. The concept of DSSCs imitating by the energy and electron transfer mechanisms in natural photosynthesis of plants and in dye sensitized silver halide emulsions used for photography. It is based on the sensitization of wide band gap n-type semiconductor through a dye used as a sensitizer. The absorption spectrum of the dye and the anchorage of the dye to the surface of titanium dioxide (TiO_2) are important parameters determining the efficiency of the cell. In other words, DSSC is a molecular machine that is one of the first devices to go beyond microelectronics into the realm of nanotechnology [2].

The history of DSSC started in 1972 when a chlorophyll sensitized zinc oxide (ZnO) electrode was developed. The first amorphous silicon solar cell in 1976 had an efficiency of 2.4%, which was later increased to 4%. In the following years, DSSC has become an attractive subject in solar energy cell research from both applied and fundamental points of view. The main dilemma was that a single layer of dye molecules on a surface enabled 1% incident sunlight absorption [3]. In 1991, Grätzel and his co-workers [4] developed a new type of solar cell, known as DSSC or Grätzel cell, having finest photovoltaic performance in terms of both conversion yield and long term stability. It mimics the photosynthesis in plants by sensitizing nano-particulate titanium dioxide (TiO_2) films with novel polypyridyl complexes of Ruthenium (Ru) dyes and found to absorb visible light up to approximately 800 nm with the energy conversion efficiency exceeding 7% [4]. Further to superior light harvesting properties and durability, a significant advantage of these dyes lies in the metal–ligand charge transfer transition through which the photoelectric charge is injected into TiO_2. In Ru complexes, this transfer takes place at a much faster rate than the back reaction, in which the electron recombines with the oxidized dye molecule rather than flowing through the circuit and performing work [5]. Metal–organic dyes (mostly Ru and osmium polypyridil complexes) have been used as efficient sensitizers. The preparation routes for metal complexes are often based on multi step reactions involving long, tedious and expensive chromatographic purification procedures. However, the use of these expensive Ru metals, derived from relatively scarce natural resource corresponds to a relatively heavy environmental burden [6]. There is currently a large effort to improve the performance of low cost renewable energy devices.

Hence, it is possible to use natural pigments/ dyes, extracted from flowers, fruit, leaves, algae, and fungi can be used as alternative photosensitizers in DSSCs with appreciable efficiencies. Their advantages over synthetic dyes include easy availability, abundance in supply, can be applied without further purification, environmentally friendly, fully biodegradable and they considerably reduce the cost of devices [7]. Also, organic dyes such as phthalocyanine, cyanine, xanthene, coumarine, etc. are also found to be poor sensitizers because of weak binding energy with TiO_2 film and low charge transfer absorption in the whole visible region [8]. Thus far, flavonoids, carotenoid, anthocyanin and betalain extracts together with selected chlorophyll derivatives are the most successful plant sensitizers.

18.2 WORKING PRINCIPLE OF DSSC

The basic principle of a DSSC with natural dyes involves exchange of the synthetic dyes (N3, N719 and 'black' dyes) of the set-up of DSSCs with natural ones. Developing and optimizing materials for organic solar cells is generally not yet rational, but rather empirical. The reason for this is that multiple parameters have to be taken into account when new types of materials and dyes are designed for organic solar cells. The photovoltaic cell's efficiency does not only depend on the pigment's molecular structure, but also strongly on solid-state properties, such as morphology, self-assembly and aggregation of the dye molecules. In the case of DSSCs even more parameters have an influence on the efficiency such as the type of the photoelectrode, the dye's anchoring group and the electrolyte with the respective redox couple. Hence, the optimal design of organic dyes regarding to an enhancement of the DSSCs' performance can only be derived from approximate trends. In general, the design of a metal-free photosensitizer consists of a donor–acceptor– substituted π-conjugated bridge for the tuning of photoelectric properties. The acceptor part is the dye's anchoring group which is attached to the TiO$_2$ (Figure 18.1) [9].

Up to now mainly electrolytes known from DSSCs with synthetic dyes are used for DSSCs, whereas DSSCs with natural dyes are new [10]. Dye-cocktails or co-sensitization could yield higher efficiencies if the absorption spectra of the different dyes do not overlap too much in order to extend the overall absorption spectrum. However, limitations could be in a lower TiO$_2$ surface area for a particular dye/pigment or intermolecular interactions between the different dyes which may result in lower electron injection efficiency [11].

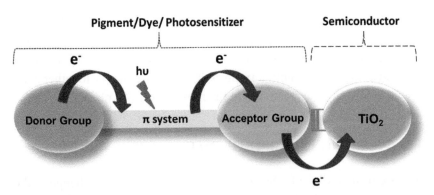

FIGURE 18.1   Donor-π-acceptor structure mechanism of an organic dye in DSSCs with TiO2 photoanodes.

The creation of a voltage difference or electric current, in a material upon exposure to light, is possible because of the photovoltaic (PV) effect, and depending on the materials. Among all, the most studied device of the last generation is the DSSC, being a low cost and high efficiency solar energy-to-electricity converter. The actual DSSCs contains broadly five components: a transparent conductive oxide (TCO) substrate (glass support), a nanostructured n-type semiconductor, a visible-light absorber dye (possess several O or –OH groups capable of chelating to the Ti (IV) sites on the TiO_2 surface), an electrolyte and a counter electrode. A schematic representation of the dye-sensitized solar cell is shown in Figure 18.2. With the key idea of mimicking the natural photosynthesis, DSSCs are being pursued as eco-friendly devices, which could be easily fabricated. A typical DSSC comprises a nano-crystalline titanium dioxide (TiO_2) electrode modified with a dye fabricated on a TCO, a platinum Pt counter electrode, and an electrolyte solution placed between the two electrodes based on the iodide/iodine redox couple (I^-/I_3^-).

In a DSSCs, upon absorption of light, the dye (S) is promoted into an electronically excited state (S*) from where it injects (within fs), an electron into the conduction band of a large band gap semiconductor film (TiO_2), onto which it is adsorbed (I). The electrons are transported through the TiO_2 film by diffusion before reaching the anode of the cell

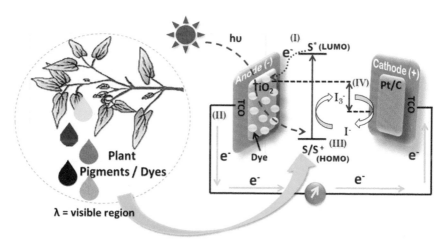

FIGURE 18.2   Schematic representation of the components and of the basic operating principle of a DSSC.

(II). The positive charges resulting from the injection process are transferred into the liquid electrolyte by the reaction of the dye cation (S+) with the reduced species of a redox couple in the electrolyte solution. This leads to the generation of the charge neutral state of the sensitizer (III). The most typical redox couple is I^-/I_3^-. After ionic diffusion, the carrier of the positive charge (I_3^-) reaches the cathode, where it releases its charge, thus being reduced back to I^- (IV) [12, 13]. Due to its multiple advantages, titanium dioxide became the semiconductor of choice for the photoelectrode. It is low cost, widely available, and non-toxic. Ruthenium complexes were employed as sensitizer very early on and are still now the most commonly used sensitizer. Finally, the main redox couple used is tri-iodide/iodine.

DSSC parallels photosynthesis in the use of a dye as the light harvester to produce excited electrons, TiO_2 replaces carbon dioxide as the electron acceptor, iodide/triiodide (I^-/I_3^-) replacing water and oxygen as the electron donor and oxidation product and a multilayer structure (similar to the thylakoid membrane) to enhance both the light absorption and electron collection efficiency. The operating cycle can be summarized in chemical reaction as [14]:

$$e^-(Pt/C) + h\upsilon \rightarrow 3I^-$$

Anode	Cathode
$S + h\upsilon \rightarrow S^*$ Absorption	
$S^* \rightarrow S^+ + (TiO_2)$ Electron injection	$I_3^- + 2e^- (Pt/C) \rightarrow 3I^-$
$2S+ + 3I^- \rightarrow 2S + I_3^-$ Regeneration	

The operation of the cell is regenerative in nature since no chemical substances are neither consumed nor produced during the working cycle, as visualized in the cell reaction.

18.3 TYPES OF PIGMENT DYES

A plant is a complex and ever active chemical factory that produces a wide range of chemical moieties that it requires for protection against yeast and molds, resistance to insect attack, even the protection against ultraviolet and drought conditions in some specialist plants. It is this complex environment that produces many dozens of chemical entities as pigments/dyes that may have benefit in human treatments for various applications.

Natural dyes provide a viable alternative to expensive organic based DSSCs. Various components of a plant have been tested over the last two decades as suitable sensitizers. Plant pigmentation occurs due to the electronic structure of the pigment interacting with sunlight to alter the wavelengths that are either transmitted or reflected by the plant tissue. The specific color will depend on the abilities of the observer. Humans without color blindness can detect wavelengths between approximately 430 and 680 nm, representing the visible spectrum. The pigments can be described in two ways: the wavelength of maximum absorbance (λmax) and the color perceived by humans. Figure 18.3 illustrates the grouping of plant pigments based on the common structure and biosynthetic basis.

18.3.1 FLAVONOID

Fruits, vegetables, fungi and some bacteria produce, as part of their secondary metabolism, in the form of a wide variety of phenolic compounds

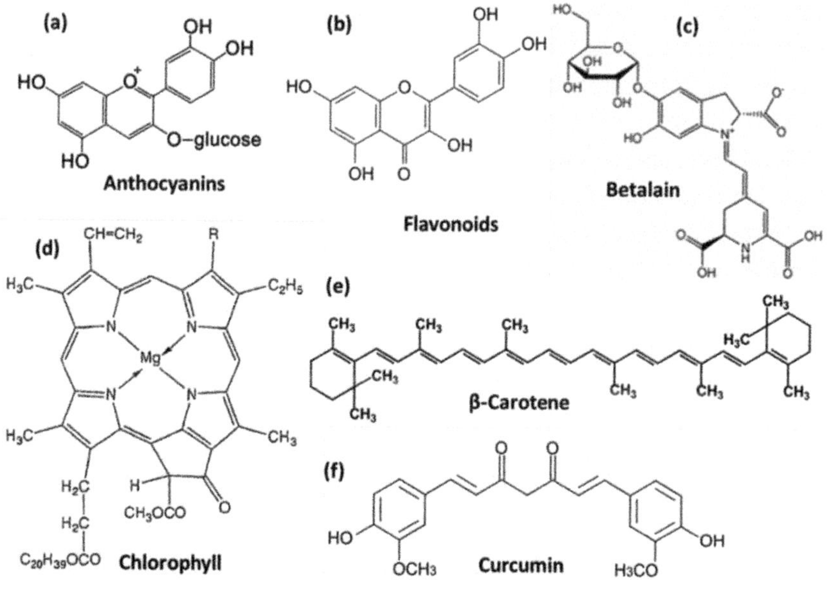

FIGURE 18.3   Chemical structure of plant pigments presents in different resources for the DSSCs.

(Ferulic Acid, Gallic acid, Ellagic Acid, Rosmarinic acid and so on). In the early 1960s, phenolic compounds were widely viewed as metabolic waste products that were stored in the plant vacuole. In foods this kind of compounds acts as pigments, antioxidants, flavor precursors, etc. and, nowadays, as part of our diet, they have been associated with several health promoting activities in the form of nutraceuticals such as: decreasing blood sugar levels, reducing body weight, anticarcinogenic, antiinflamatory, anti-aging and antithrombotic activity [15]. *Flava* means yellow in Greek and the collective name of flavonoids for this group of compounds was proposed by Geissman in 1952. This is a very large group of natural products, showing extraordinary diversity and variation and as the Greek root for the group suggests, as many of these compounds are yellow in color. One of the major phenolic phytochemicals is flavonoids, which are basically comprised of fifteen carbons, with two aromatic rings conjugated by a three-carbon bridge (C_6–C_3–C_6 structure) that usually forms a third ring (Figure 18.3b). The flavonoids (Quercetin and its glucosides, Apigenin, Genistein, Luteolin, etc.) are a group of low-molecular-weight polyphenolic substances that are formed from the combination of derivatives synthesized from phenylalanine (via the shikimic acid pathway) and acetic acid [16]. Lowest intake (1 to 9 mg/d) of flavonoids was from a South American diet, whereas highest flavonoid intake (75 to 81 mg/d) was from a Scandinavian diet [17].

Several biological activities have been attributed to flavonoids such as antioxidative, anticarcinogenic, vasoprotective, anti-inflammatory, neuro-degenerative, antidiabetic, and antiplatelet activity [18].

Natural products such as chalcones and aurones also contain a C_6–C_3–C_6 backbone and are considered as minor flavonoids. The various colors of flavonoids are determined by the degree of oxidation of the C-ring. Despite the similarities in structure, only some flavonoids have the ability to absorb light in the visible region of the spectrum and are thus pigments. Pigment molecules are characterized by having unbound or loosely bound electrons. For such molecules, the energy required for excitation of the electrons to a higher energy level is lowered, allowing the molecule to be energized by light within the visible range. The color of the pigment is determined by the particular wavelengths of visible light that are absorbed by the molecule and those by the particular wavelengths of visible light that are absorbed by the molecule and those that are reflected or scattered. The structural features of a flavonoid pigment are the degrees of double bond conjugation and oxygenation (in the form of hydroxylation).

For many angiosperms, color is the key to attracting pollinators, whether they are bees, butterflies, other insects or birds, although it is frequently one of a number of factors including fragrance, floral shape and nectar reward. In plants, flavonoids have several functions including attracting insects for pollination and dispersal of seeds, acting in defense systems such as UV-B protectants and phytoalexins, signaling between plants and microbes and regulating auxin transport. Flavonoids may form UV–visible patterning in petals, often in combination with UV-reflective carotenoid pigments. While in most flowers, the color begins with the production and accumulation of flavonoid chromophores, other factors both intrinsic and extrinsic, come into play that determine the actual color that is manifested by the flower. Flavonoid pigmentation is basically based on three components: the primary structures of the flavonoids, secondary structures of these molecules due to pH and tertiary structures arising from self-association and inter and intramolecular interactions [19, 20]. However, the major claimed activity of phenolic compounds and flavonoids has been seen as antioxidants whereas, their importance of color have been significantly attracted in DSSCs industry.

18.3.2   ANTHOCYANINS

The anthocyanins constitute a major flavonoid group that is responsible for cyanic colors ranging from salmon pink through red and violet to dark blue of most flowers, fruits and leaves of angiosperms. Sometimes they are present in other plant tissues such as roots, tubers and stems [21]. It does not show their brilliant colors until they are accumulated in the acidic vacuoles or spherical vesicles called 'anthocyanoplasts.' Recently, it has considerable interest within the food industry because of their potential application as natural colorants, nutraceuticals, antioxidants and chronic diseases (such as heart disease, cancer, and inflammation) prevention. However, the commercial application of anthocyanins as colorants and nutraceuticals in processed foods is often limited because of their poor chemical stability, which depends on their chemical structure, concentration, pH, temperature, oxygen, light, polymeric forms, and the presence of cofactors and/or ascorbic acid [22]. Some flavonols have a protective role as UV-B filters and they also could function as copigments for anthocyanins for special tissues [20]. The main sources of anthocyanins are some vegetables such as blackberries, grapes, blueberries, eggplant (aubergine), red cabbage, red radicchio, red oranges, elderberry, red onion, black rice, mango and purple corn. The most common anthocyanidins found in flowers are pelagonidin (orange), cyanidin (orange-red), delpinidin (blue-red), petunidin (blue-red) and malvidin (blue-red). They are glycosylated polyhyroxy or polymethoxy derivatives of 2-phenylbenzopyrylium or flavylium salts, which consist of three six-membered rings [23], as shown in Figure 18.3a.

18.3.3   CAROTENOIDS

Carotenoids are C_{40} tetraterpenoids natural pigments biosynthesized by the linkage of two C_{20} Geranylgeranyl diphosphate molecule presents in plants, algae, yeast, fungi and photosynthetic bacteria; and responsible for the red, orange, and yellow colors in various fruits and vegetables. They are usually responsible for petal colors in yellow-to-orange and prominent for their distribution, structural diversity and various functions. Fruits and vegetables provide most of the carotenoids in the human diet. Carotenoids can be broadly classified into two classes, carotenes (α-carotene, β-carotene or lycopene) (Figure 18.3e) and xanthophylls (β-cryptoxanthin, lutein or zeaxanthin) [24]. The range of colors is expanded by diverse modifications to the simple polyene chain structure. The carotenoid backbone is either linear or contains cyclic end groups and the two major classes constitute of carotenes and their oxygenated derivatives, the xanthophylls. Carotenoids are involved in photosystem assembly and contribute to light harvesting by absorbing light energy in a region of the visible spectrum where chlorophyll absorption is lower and by transferring the absorbed energy to chlorophyll. It also provides protection from excess light via energy dissipation, free radical detoxification and limits damage to membranes [25].

18.3.4   CURCUMIN

Curcumin, which is derived from the ground rhizome of *Curcuma longa* L. (turmeric), native to tropical South Asia and needs temperatures between 20°C and 30°C (68°F and 86°F) and a considerable amount of annual rainfall to thrive. When not used fresh, the rhizomes are boiled for several hours and then dried in hot ovens, after which they are ground into a deep orange-yellow powder commonly used as a spice in curries and other South Asian and Middle Eastern cuisine, for dyeing, and to impart color to mustard condiments [26]. Curcumin (Figure 18.3f) can be used to test the alkalinity or acidity of foods. It turns yellow in an acidic food, and it turns red in an alkaline food. It is also well-known for its pharmaceutical applications had not gained much insight as a photosensitizer, in spite of its photoactivity in the visible region of the solar spectrum. These curcumin-derived dyes are showing intense, long wavelength absorption in the visible region of 420–580 nm, however, till date, their implication as a sensitizer in DSSCs is nascent. Curcumin has high thermal and chemical stability, eco-friendly, along with low production cost that would be certainly interesting as a naturally derived dye sensitizer in DSSCs applications [27].

18.3.5   BETALIN

Betalains are a class of red and yellow indole-derived pigments (Figure 18.3c), which are found in plants of the order *Caryophyllales*. The most promising source of betalains pigments is the prickly pear fruit, beet-root, bougainvillea, amaranth, and other species, which found widely throughout the world [28]. The betalain pigments comprise the red-purple betacyanins, betanin (I) and betanidin (II), with maximum absorptivity at λmax about 535 nm, and the yellow betaxanthins with λmax near 480 nm. Betanin, the red-purple pigment distributed in beets, is the 5-O-β-glucoside of betanidin. The aglycone betanidin has a slightly red-shifted absorption spectrum compared to betanin. Interest in betalain pigments for their potential applications as natural food colors, antioxidants, and light absorbing properties, are capable of complexing metal ions, and exist in nature in association with various copigments which modify their light absorption properties [29].

18.3.6 CHLOROPHYLL

Leaves of most plants are rich in chlorophyll (Chl) and its application as a natural dye sensitizer has been experimented in many associated studies. Chlorophyll (Figure 18.3d) is a green pigment found in the leaves of most green plants, algae, and cyanobacteria, responsible for the absorption of energy from light during the photosynthesis process where, the light absorbed by chlorophyll transforms carbon dioxide and water into carbohydrates and oxygen [$CO_2 + H_2O \rightarrow (CH_2O) + O_2$]. Six different types of Chl pigment exist, and the most occurring type is Chl a. The molecular structure includes a chlorine ring with Mg center, along with different side chains and a hydrocarbon trail, depending on the Chl type. Chl absorbs light from red, blue, and violet wavelengths, and derives its color by reflecting green. Chls are the principal pigments in natural photosynthetic systems [30]. Their functions include harvesting sunlight, converting solar energy (to chemical energy), and transferring electrons. Chls include a group of more than 50 tetrapyrrolic pigments [31]. Chls and their derivatives are inserted into DSSC as dye sensitizers because of their beneficial light absorption tendency modes; the most efficient of which is Chl a (chlorine 2) derivative-methyl trans-32-carboxy-pyropheophorbide. Chl has an absorption maximum at 670 nm because of an attractive compound that acts as a photosensitizer in the visible light range and do not contain a heavy metal ion, and are thus suitable photosensitizers from the environmental viewpoint [32].

18.4   SOURCE OF PIGMENT DYES FOR DSSCS

Nanocrystalline DSSCs are promising synthetic nanomachines based on principles similar to the processes in natural photosynthesis. Both use an organic dye to absorb the incoming light and produce excited electrons. A film of interconnected nanometer-sized TiO_2 particles replace nicotinamide adenine dinucleotide phosphate ($NADP^+$) and carbon dioxide as the electron acceptor, and iodide and triiodide (I^-, I_3^-) replace water and oxygen as the electron donor and oxidation product, respectively [33]. Theoretical achievements of the chemical and physical properties of natural dye as sensitizers are important in detecting the relationship between the performance and structure properties of dyes, as well as in the design and synthesis of new dye sensitizers. The main factor for the increased cell performance is the reduced interaction among the excited state dye sensitizers [34]. Therefore, natural pigments present in the flowers, fruits, leaves, roots and so on may be used as natural water-based substitutes for the ruthenium bipyridyl dyes. Photoelectrochemical parameters of natural dye based DSSCs have been shown in Table 18.1.

18.5 PHOTOELECTROCHEMICAL PARAMETERS

The photoelectrochemical parameters of DSSCs using a range of natural dyes as photosensitizers subject to the work done by various pioneers. It demonstrates typical characteristics such as η, J_{sc}, V_{oc} and fill factor (FF) values. The FF is a measure of the junction quality and the series resistance of the DSSC and is typically calculated as follows:

$$FF = (J_m \times V_m)/J_{sc} \times V_{oc} \tag{1}$$

where, J_m and V_m are the maximum current and voltage, respectively.

Solar energy-to-electricity conversion efficiency, η, under sunlight irradiation (e.g., AM 1.5) can be obtained as follows:

$$\eta = (J_{sc} \times V_{oc} \times FF)/P_{in} \tag{2}$$

where, P_{in} is the power of incident light and measured in $mWcm^{-2}$ using a solarimeter; and J_{sc} is measured in $mAcm^{-2}$ [12, 38].

18.6 IMPORTANCE OF FUNCTIONAL GROUPS OF DYES FOR DSSCS

Based on the results with Ru-complexes also a variety of metal– organic dyes such as Zn-porphyrin derivatives have been tested [39]. Prerequisites

TABLE 18.1   Photoelectrochemical Parameters of Natural Dye Based DSSCs [35]

Plant materials	Dye/Pigment	λmax (nm)	J_{sc} (mA cm^{-2})	V_{oc} (V)	FF & η (%)	Ref.
Jathopha curcas Linn (Botuje)	Flavonoid	400	0.69	0.05	0.87/0.12	–
Punica granatum peel	Flavonoid	400	3.341	0.716	0.776/1.86	–
Cosmos sulfureus	Flavonoid	505, 590	1.041	0.447	0.61/0.54	–
Tangerine peel	Flavone	446	0.74	0.592	63.1/0.28	–
Mangosteen pericarp	Rutin	–	2.69	0.686	63.3/1.17	–
Rhododendron	Anthocyanin	540	1.61	0.585	60.9/0.57	–
Begonia	Anthocyanin	540	0.63	0.537	72.2/0.24	–
Bauhinia tree	Anthocyanin	665	0.96	0.572	66.0/0.36	–
Rosa xanthina	Anthocyanin	560	0.64	0.49	0.52	–
Daucus carota L (Black carrot)	Anthocyanin	540	1.304	0.4	0.47/0.25	–
Hibiscus rosasi-nesis	Anthocyanin	534	4.04	0.40	0.63/1.02	–
Punica granatum (Pomegranate)	Anthocyanin	412, 665	2.05	0.56	0.52/0.59	–
Ixora macro-thyrsa	Anthocyanin	537	1.31	0.40	0.57/0.30	–
Nerium olender	Anthocyanin	539	2.46	0.40	0.59/0.59	–
Sesbania grandi-flora	Anthocyanin	544	4.40	0.41	0.57/1.02	–
Allium cepa (Red onion)	Anthocyanin	532	0.51	0.44	0.48/0.14	–
Lawsonia inermis (Henna)	Anthocyanin	518	1.87	0.61	0.58/0.66	–
Myrtus cauliflora mart (Jaboticaba)	Anthocyanin	520	7.20	0.59	0.54	–
Kopsia flavida	Anthocyanin	550	1.20	0.52	0.62	–
Lithospermum	Anthocyanin	520	0.14	0.337	58.5/0.03	–
Rose	Anthocyanin	–	0.97	0.595	65.9/0.38	–
Hibiscus sabdar-iffa L. (Rosella)	Anthocyanin	520	1.63	0.40	0.57/0.37	–
Blue pea flower	Anthocyanin	580	0.37	0.37	0.33/0.05	–

TABLE 18.1 (Continued)

Plant materials	Dye/Pigment	λmax (nm)	J_{sc} (mA cm^{-2})	V_{oc} (V)	FF & η (%)	Ref.
Hibiscus surattensis	Anthocyanin	545	5.45	0.39	0.54/1.14	–
Solanum melongena (Eggplant)	Anthocyanin	522	3.40	0.35	0.40	–
Mulberry	Anthocyanin	543	0.86	0.42	0.43	–
Hylocereus costaricensis (Dragon fruit)	Anthocyanin	535	0.20	0.22	0.30/0.22	–
Brassica olercea *(Redcabbage)*	Anthocyanin	537	0.50	0.37	0.54/0.13	–
Citrus sinensis (Redsicilian)	Anthocyanin	515	3.84	0.34	0.50	–
Nephelium lappaceum (Rambutan)	Anthocyanin	540	3.88	0.404	0.35/0.56	–
Chaste tree fruit	Anthocyanin	548	1.06	0.39	0.48	–
Oryza sativa L. indica (Black Rice)	Anthocyanin	560	1.14	0.55	68.7/0.33	–
Cheries	Betacyanin	500	0.46	0.30	38.3/0.18	–
Grapes	Betacyanin	560	0.09	0.34	61.1/0.38	–
Raspberries	Betacyanin	540	0.26	0.42	64.8/1.50	–
Violet Bougainvillea glabra	Betacyanin, Betaxanthin	547	1.86	0.23	0.71/0.31	–
Blood leaf	Betacyanin	460	0.260	0.267	0.46/0.04	–
Fructus lycii	Carotene	447, 425	0.53	0.689	46.6/0.17	–
Ivy gourd fruits	β-Carotene	458, 480	0.24	0.64	0.49/0.09	–
Bixa arellana L. (Achiote)	Carotenoid	474	1.1	0.57	0.59/0.37	–
Chlorella sp. PP1	Carotenoid					–
Allamanda cathartic (Golden trumpet)	Carotenoid	325, 458	0.878	0.405	0.54/0.40	–
Capsicum	Carotenoid	455	0.23	0.41	0.63/–	–
Walnuts	Carotenoid	510	0.73	0.304	0.39/0.1	–

TABLE 18.1   (Continued)

Plant materials	Dye/Pigment	λmax (nm)	J_{sc} (mA cm^{-2})	V_{oc} (V)	FF & η (%)	Ref.
Gardenia yellow	Carotene	450	0.96	0.54	0.62/0.32	–
Marigold flower	Xanthophyll	487	0.51	0.542	83.1/0.23	–
Yellow rose flower	Xanthophyll	487	0.74	0.609	57.1/0.26	–
Tagetes erecta	Xanthophyll	465	2.89	0.475	0.606/0.8	–
Turmeric	Curcumin	525	0.288	0.529	0.48/0.03	–
Bougainvillea	Betalin	300	2.10	0.3	–/0.36	–
Red turnip	Betalin	536	9.50	0.45	0.37/1.7	–
Wild sicilian prickly pear	Betalin	465	8.20	0.38	0.38/1.19	–
Erythrina varie-gata	Chlorophyll	451, 492	0.78	0.48	0.55/–	–
Perilla	Chlorophyll	665	1.36	0.522	69.9/.050	–
Anethum graveo-lens	Chlorophyll	666	0.96	0.57	40.0/0.22	–
Spinach	Chlorophyll	437	0.47	0.55	0.51/0.13	–
Ipomea	Chlorophyll	410	0.91	0.54	0.56/0.28	–
China Loropetal	Chlorophyll	665	.084	0.518	62.6/0.27	–
Petunia	Chlorophyll	665	0.85	0.616	60.5/0.32	–
Lemon leaves	Chlorophyll	475, 675	0.286	0.539	0.74/0.05	–
Herba artimisiae scopariae	Chlorophyll	669	1.03	0.484	68.2/0.34	–
Chinese rose	Chlorophyll	516	0.90	0.483	61.9/0.27	–
Vernonia amygda-lin (Bitterleaf)	Chlorophyll	400	0.07	0.34	0.81/0.69	–
Rhubarb	Chlorophyll	410	0.82	0.238	0.44/0.01	–
Petroselinum crispum (Parsley)	Chlorophyll	666	0.53	0.44	34.0/0.07	–
Rhoeo spathacea	Chlorophyll	670	10.9	0.50	0.37/1,49	–
Sargassum wightii (marine seaweed)	Chlorophylls, carotenoids, fucoxanthin	412.5, 610, 659.5	0.203	0.33	0.46/0.07	36
Cochineal	Carminic acid	515, 544	0.51	0.78	0.25/0.10	37

for photosensitizers to function in DSSCs are the absorption in the visible or near-infrared regions of the solar spectrum and the binding to the semiconductor TiO_2 [40]. However, the DSSCs sensitized by the natural dyes used in this work did not offer high conversion efficiency. This can be explained as due to the lack of available bonds between the dye and TiO_2 molecules which electrons can transport from the excited dye molecules to the TiO_2 film. The functional group necessary to interact with the TiO_2 surface is a carboxylic or other peripheral acidic anchoring group [41]. The TiO_2 binding moiety of large π-aromatic molecules is very often a carboxylic group [42].

There are several possible chemical functional groups that are able to bind the dye to the TiO_2. The best anchoring groups for metal oxides are phosphonic acids followed by carboxylic acids and their derivatives, such as acid chlorides, amides, esters or carboxylate salts [41]. The carboxylic group has been the most frequently used anchoring groups. In order to form bonds, the dye's binding groups react with surface hydroxyl groups (-OH) of the metal oxide. In the case of phosphonic acids or carboxylic acids, a reversible binding with high equilibrium binding constants is established between the photosensitizer and TiO_2. In basic conditions (usually pH > 9) the dyes are easily desorbed again. There are several binding modes between the TiO_2 and a dye molecule with at least one carboxylic group (COOH). Which mode of chemisorption between TiO_2 and COOH is prevalent depends on the dye's structure, its anchoring groups, the pH, and the metal oxide's preparation [43]. The interaction of TiO_2 with cyanin, an anthocyanidine, is via the hydroxyl-ketone tautomer of cyanin [43].

The main issues which have to be taken into account are an appropriate anchoring of the molecules, while allowing a fast and energetically optimized charge transfer. The π electrons of the carboxylic group should be in resonance with the π electrons of the polyene backbone in order to be able to shift the electrons through the anchoring group into the conduction band of the semiconductor. The color of dyes also depends on such delocalized π electrons. This result indicates that the interaction between the sensitizer and the TiO_2 film is significant in enhancing the energy conversion efficiency of DSSCs (Figure 18.4).

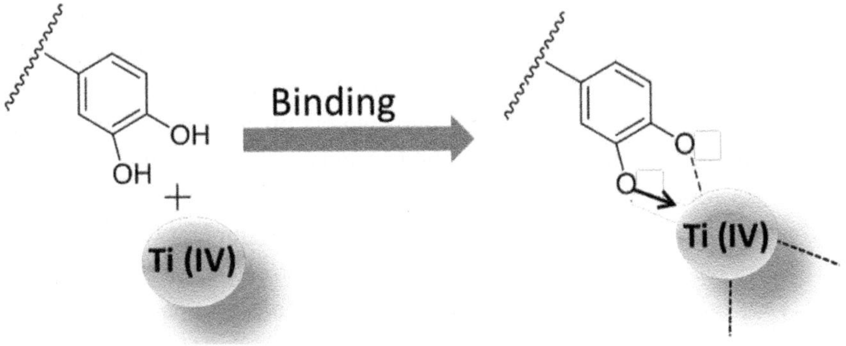

FIGURE 18.4   Pigment-Ti(IV) complex formed via the adsorption of the dye (-OH) from solution onto the titanium dioxide surface.

18.7   EFFICIENCY AND DRAWBACKS

The stability of the four components of a DSSC, namely, dye sensitizer, a counter electrode, electrolyte, and semiconducting oxide, has been subjected to close scanning. An ideal sensitizer for DSSC that is capable of converting standard global sunlight to electricity should absorb all the sunlight below the threshold wavelength of about 920 nm (1.4 eV). Thus, continuing scientific research that aims to further enhance the efficiency and long-term stability is required to fulfill the promises of future technology. A DSSC device must remain applicable for 20 years without performance loss. The photostability of the DSSC sensitizer material must be capable of many redox cycles without decomposition, and must also have the ability to carry attachment groups, such as phosphonate or carboxylate, to promptly graft it to TiO_2 oxide [44].

The dye attachment group must spontaneously form a layer upon the oxide film. High probability dispersion of this molecular dye layer occurs upon photon absorption, and the relaxation of the dye molecule occurs when the excited electron is injected into the mesoporous conduction band. However, single dye monolayer absorption is weak; thus, producing a highly efficient sensitizing device is impossible, assuming that smooth substrate surfaces are imperative to avoid the recombination loss mechanism associated with polycrystalline (rough) structures in the solid state photovoltaic. The barrier for charge recombination exists when the electron produced in semiconductor lattice is separated from the positive charge carriers by the dye molecule layer, which is an insulator in the ground state. Utilizing a roughness factor of over 1000 with nanocrystalline thin film has become the standard practice [45].

Limited long-term stability, difficulty in hermetic sealing and leakage of the liquid electrolyte have been identified as relevant drawbacks in view of real indoor and outdoor applications of DSSCs. A possible solution is represented by the entrapment of the liquid electrolyte by means of a polymeric or inorganic network. Polymeric quasi-solid electrolytes can be prepared in the form of gels or membranes where a huge amount of cross-linked matrices has been investigated. As an alternative, inorganic nanoparticles (NPs) have been adopted to jell liquid electrolytes, creating a quasi-solid (paste-like) network containing self-assembled channels able to ensure sufficient ionic transport. To date, quasi-solid electrolytes are considered the optimum compromise between efficiency and durability, and the recent advances in this subject have been collected in some review articles [13, 46–48].

The rest of the natural dye based DSSC produced efficiencies less than 1%, ranging from 0.05% to 0.59% as illustrated in Table 18.1. The reason for such low efficiency is that the interaction between pigments and TiO_2 would have been low. The structure of the pigment also affects the performance, i.e., if the structure has a longer R group, this result in the steric hindrance for the pigment to form bond with the oxide surface of the TiO_2 and hence prevents the molecule from arraying on the TiO_2 film effectively. Hence there is lack of electron transfer from the dye molecules to the conduction band of TiO_2. The intensity and range of light absorption of the extract also affect the performance. The distance between the dye skeleton and the point connected to the TiO_2 surface facilitates electron transfer from the dye molecule to the TiO_2 surface which accounts for the performance of the cell [49].

The interaction between TiO_2 and the dye plays an important role towards the efficiency of DSSCs. In general, natural dyes suffer from low V_{oc}. According to Calogero et al. [50], this can be due to possible inefficient electron/dye cation recombination pathways and the acidic dye adsorption environment. In fact H^+ are the potential determining ions for TiO_2 and that proton adsorption causes a positive shift of the Fermi level of the TiO_2, thus limiting the maximum photovoltage that could be delivered by the cells. According to, the charge transfer in the TiO_2/dye/electrolyte interface resistance leads to a decrease in Jsc. Thus, introducing a functional group, such as carboxyl group and optimizing the structure of the natural dye are necessary to improve the efficiency of natural dye based DSSCs.

Some complication such as dye aggregation on nanocrystalline film produces absorptivity that results in no electron injection. Dye aggregation is a serious issue that occurs when compounds fill the free space between the dye molecules, partially blocking the physical contact between the iodine solution and TiO_2 semiconductor film surface, reducing reaction and inhibiting dye aggregation [51].

18.8   RECOMMENDATIONS

Many factors affect the performance of DSSC. Some parameters include the nature of the pigment, light spectrum and intensity, the thickness of the TiO_2 layer, the size of the active area, the nature of the electrolyte salt and the extracting solvent. For an efficient solar cell, rapid charge injection and transport and intense visible light absorption are necessary. Although the efficiencies obtained with these natural dyes are still below the present conditions for large-scale practical application, the results are encouraging and may boost additional studies oriented to the search of new natural sensitizers and optimization of the solar cell [50].

According to Cherepy et al. [52], characterization of the loss of carriers due to electron recapture by the electrolyte as the electrons drift through the film to the back contact is important for future research. Also finding various additives to improve V_{oc} can result in an increase in energy conversion. The dye can be affected by oxygen, pH, temperature and other variables that lead to its degradation and affect the performance of DSSCs [53]. Hence, further investigations can be performed on this subject. Solid-state electrolytes are possible replacements to the liquid ones, however the efficiency becomes low. Hence a possible future research could be to find ways of improving its performance. Stability of natural dye based DSSC is also a major concern for the natural dye based DSSC. The implication of decrease in stability makes the entry of this device into practical applications, difficult. Therefore, further studies must be performed to improve the stability of the cell.

18.9 CONCLUSIONS

DSSC represents the most widely studied device for the conversion of solar energy into electricity and plant pigments/dyes as sensitizers for DSSCs are promising alternative to organic dyes. They offer environmental friendliness, low-cost production, designable polychrome modules, simple preparation technique and wide availability. Despite the efficiencies obtained with the natural dyes are below the requirements for large-scale production, the results are hopeful and can boost additional studies oriented to the search of new natural sensitizers and to the optimization of solar cell components compatible with such dyes. The use of water in solar energy conversion devices was intended to be a big challenge by a great part of the scientific community. Thus, the utilization of food and non-food based crops in industrial processes for the production of DSSCs, could be an additional source of revenue for farmers and also help in agro-industry diversification by providing a platform for dyes.

18.10 SUMMARY

A wide variety of pigments obtained from natural sources such as plants, insects/animals and microbes with eco-friendly, robust and cost effective processing technologies. Natural dyes have become a viable alternative to expensive and rare organic sensitizers because of its low cost, easy attainability, abundance in supply of raw materials and no environmental threat. The nature of these pigments together with other parameters has resulted in varying performance. This book Chapter briefly describes the correlation between natural pigments and DSSCs used over the years.

ACKNOWLEDGMENTS

This scientific work has been funded by the Prometeo Project of the Ministry of Higher Education, Science, Technology and Innovation (SENES-

CYT), Ecuador. The author is also grateful to the TATA College for the financial support.

KEYWORDS

- dye-sensitized solar cells
- efficiency
- environmental threat
- low cost
- natural dyes
- natural pigments
- organic sensitizers
- solar cells
- solar light

REFERENCES

1. Calogero, G., Bartolotta, A., Marco, G. D., Carlo, A. D., & Francesco Bonaccorso (2014). Vegetable-based dye-sensitized solar cells. Chem. Soc. Rev., 44, 3244–3294.
2. Smestad, G. P. (1998). Education and solar conversion: demonstrating electron transfer. Solar Energy Materials and Solar Cells, 55, 157–178.
3. Ludin, N. A., Al-Alwani Mahmoud, A. M., Mohammad, A. B., Kadhum, A. A. H., Sopian, K., & Karim, N. S. A. (2014). Review on the development of natural dye photosensitizer for dye-sensitized solar cells. Renewable and Sustainable Energy Reviews, 31, 386–396.
4. O'Regan, B., & Grätzel, M. (1991). A low-cost, high-efficiency solar-cell based on dye sensitized colloidal TiO2 films. Nature, 353, 737–740.
5. Ryan, M. (2009). Progress in ruthenium complexes for dye sensitized solar cells. Platinum Metals Review, 53, 216–218.
6. Reijnders, L. (2010). Design issues for improved environmental performance of dye sensitized and organic nanoparticulate solar cells. Journal of Cleaner Production, 18, 4307–4312.
7. Wongcharee, K., Meeyoo, V., & Chavadej, S. (2007). Dye-sensitized solar cell using natural dyes extracted from rosella and blue pea flowers. Solar Energy Materials and Solar Cells, 91, 566–571.

8. Hao, S., Wu, J., Huang, Y., & Lin, J. (2006). Natural dyes as photosensitizers for dyesensitized solar cell. Sol Energy, 80, 209–214.

9. Mishra, A., Fischer, M., & Bäuerle, P. (2009). Metal-free organic dyes for dye-sensitized solar cells: from structure: property relationships to design rules. Angew Chem Int Edit, 48, 2474–2499.

10. Calogero, G., Yum, J., Sinopoli, A., Di Marco, G., Grätzel, M., Nazeeruddin, & M. K. (2012). Anthocyanins and betalains as light-harvesting pigments for dye-sensitized solar cells. Sol. Energy, 86, 1563–1575.

11. Kimura. M., Nomoto, H., Masaki, N., & Mori, S. (2012). Dye molecules for simple cosensitization process: fabrication of mixed-dye-sensitized solar cells. Angew Chem Int Edit, 51, 4371–4374.

12. Narayan, M. R. (2012). Review: Dye sensitized solar cells based on natural photosensitizers. Renewable and Sustainable Energy Reviews, 16, 208–215.

13. Bella, F., Gerbaldi, C., Barolo, C., & Gratzel, M. (2015). Aqueous dye-sensitized solar cells. Chem. Soc. Rev., 44, 3431–3473.

14. Matthews, D., Infelta, P., & Grätzel, M. (1996). Calculation of the photocurrent-potential characteristic for regenerative, sensitized semiconductor electrodes. Solar Energy Materials and Solar Cells, 44, 119–55.

15. Bernal, J., Mendiola, J. A., Ibáñez, E., & Cifuentes, A. (2011). Advanced analysis of nutraceuticals. Journal of Pharmaceutical and Biomedical Analysis. 55, 758–774.

16. Heller. W. (1986). Flavonoid biosynthesis: an overview. In: Plant Flavonoids In Biology and Medicine: Biochemical, Pharmacological, and Structure–Activity Relationships. Cody, V., Middleton, E., & Harborne, J. B. (eds.), pp. 25–42, New York, NY, USA.

17. de Vrie, J. H. M., Karin Janssena, P. L. T. M., Hollman, P. C. H., van Staverena, and, W. A., Katan, M. B. (1997). Consumption of quercetin and kaempferol in free-living subjects eating a variety of diets. Cancer Letters, 114, 141–144.

18. Survay, N. S., Upadhyaya, C. P., Kumar, B., Ko, E. Y., & Park, S. W. (2011). New Genera of Flavonols and Flavonol Derivatives as Therapeutic Molecules, Journal of the Korean Society for Applied Biological Chemistry, 54(1), 1–18.

19. Davies, K. (2004). Plant Pigments and Their Manipulation. USA: Blackwell Publishing Ltd.

20. Grotewold, E. (2006). The Science of Flavonoids. Columbus, USA: Springer.

21. Andersen, Q. M., & Markham, K. R. (2006). Flavonoids chemistry, biochemistry and application. NY, USA: CRC Press Taylor & Francis Group.

22. Arroyo-Maya, I. J., & McClements, D. J. (2015). Biopolymer nanoparticles as potential delivery systems for anthocyanins: Fabrication and properties, Food Research International 69, 1–8.

23. Fernando, J. M. R. V., & Senadeera, G. K. R. (2008). Natural anthocyanins as photosensitizers for dye-sensitized solar devices. Current Science, 95, 663–666.

24. Namitha, K. K., & Negi, P. S. (2010). Chemistry and biotechnology of carotenoids. Critical Reviews in Food Science and Nutrition, 50(8), 728–760.

25. Davies, K. (2004). Plant Pigments and Their Manipulation. USA: Blackwell Publishing Ltd.

26. Chan, E. W. C., Lim, Y. Y., Wong, S. K., Lim, K. K., Tan, S. P., Lianto, F. S., & Yong, M. Y. (2009). Effects of different drying methods on the antioxidant properties of leaves and tea of ginger species. Food Chemistry, 113 (1), 166–172.

27. Ganesh, T., Kim, J. H., Yoon, S. J., Kil, B.-H., Maldar, N. N., Han, J. W., & Han, S. H. (2010). Photoactive curcumin-derived dyes with surface anchoring moieties used in ZnO nanoparticle-based dye-sensitized solar cells. Materials Chemistry and Physics, 123, 62–66.

28. Calogero, G., Yum, J.-H., Sinopoli, A., Marco, G. D., Gratzel, M., & Nazeeruddin, M. K. (2012). Anthocyanins and betalains as light-harvesting pigments for dye-sensitized solar cells, Solar Energy, 86, 1563–1575.

29. Zhang, O., Lanier, S. M., Downing, J. A., Avent, J. L., Lum, J., & McHale, J. L. (2008). Betalain pigments for dye-sensitized solar cells, Journal of Photochemistry and Photobiology A: Chemistry, 195, 72–80.

30. Chang, H., Kao, M. J., Chen, T. L., Kuo, H. G., Choand, K. C., & Lin, X.-P. (2011). Natural sensitizer for dye-sensitized solar cells using three layers of photoelectrode thin films with a Schottky barrier. Am J Eng Appl Sci, 4, 214–222.

31. Scheer, H. I. In: Light-Harvesting Antennas in Photosynthesis. Green, B. R., Parson, W. W., (ed.). Dordrecht: Kluwer Academic Publishers, 2003, p. 513

32. Ludin, N. A., Al-Alwani Mahmoud, A. M., Mohamad, A. B., Kadhum, A. A. H., Sopian, K., & Karim, N. S. A. (2014). Review on the development of natural dye photosensitizer for dye-sensitized solar cells. Renewable and Sustainable Energy Reviews. 31, 386–396.

33. Smestad, G. P. (1998). Demonstrating Electron Transfer and Nanotechnology: A Natural Dye–Sensitized Nanocrystalline Energy Converter. Journal of Chemical Education 75(6), 752–756.

34. Zhang, C. R., Liu, Z. J., Chen, Y. H., Chen, H. S., Wu, Y. Z., & Yuan, L. H. (2009). DFT and TDDFT study on organic dye sensitizers D5, DST and DSS for solar cells. Journal of Molecular Structure: Theochem, 899, 86–93.

35. Shalini, S., Balasundara Prabhu, R., Prasanna, S., Mallick, T. K., & Senthilarasu, S. (2015). Review on natural dye sensitized solar cells: Operation, materials and methods. Renewable and Sustainable Energy Reviews 51, 1306–1325.

36. Anand, M., & Suresh, S. (2015). Marine seaweed Sargassum wightii extract as a low-cost sensitizer for ZnO photoanode based dye-sensitized solar cell. Adv. Nat. Sci.: Nanosci. Nanotechnol. 6, 035008 (8 pp.).

37. Park, K.-H., Kim, T.-Y., Han, S., Ko, H. S., Lee, S. H., Song, Y. M., Kim, J. H., & Lee, J. W. (2014). Light harvesting over a wide range of wavelength using natural dyes of gardenia and cochineal for dye-sensitized solar cells. Spectrochimica Acta Part A: Molecular and Biomolecular Spectroscopy 128, 868–873.

38. Wenham, S. R., Green, M. A., Watt, M. E., & Corkish, R. (2006). Applied photovoltaics. Australia: Centre for Photovoltaic Engineering.

39. Willinger, K., & Thelakkat, M. (2012). Photosensitizers in solar energy conversion. In: Photosensitizers in Medicine, Environment, and Security. Nyokong, T., Ahsen, V., (ed.). Netherlands: Springer. pp. 527–617.

40. Cherepy, N. J., Smestad, G. P., Grätzel, M., & Zhang, J. Z. (1997). Ultrafast electron injection: implications for a photoelectrochemical cell utilizing an anthocyanin dye-sensitized TiO2 nanocrystalline electrode. J Phys Chem B, 101, 9342–9351.

41. Galoppini, E. (2004). Linkers for anchoring sensitizers to semiconductor nanoparticles. Coord Chem Rev, 248, 1283–1297.

42. Imahori, H., Umeyamam, T., & Itom, S. (2009). Large π-aromatic molecules as potential sensitizers for highly efficient dye-sensitized solar cells. Acc Chem Res, 42, 1809–1818.

43. Hug, H., Bader, M., Mair, P., & Glatzel, T. (2014). Biophotovoltaics: Natural pigments in dye-sensitized solar cells, Applied Energy, 115, 216–225.

44. Grisorio, R., Luisa, D. M., Giovanni, A., Roberto, G., Gian, P. S., Michele, et al. (2013). Anchoring stability and photovoltaic properties of new D(-π-A)2 dyes for dye-sensitized solar cell applications. Dyes Pigments 98, 221–231.

45. Grätzel, M., Dye sensitized solar cells. J. Photochem. Photobiol. C. Photochem. Rev. 2003, 4, 145–153.

46. Jiang, N., Sumitomo, T., Lee, T., Pellaroque, A., Bellon, O., Milliken, D., & Desilvestro, H. (2013). High temperature stability of dye solar cells, Sol. Energy Mater. Sol. Cells, 119, 36–50.

47. Bella, F., Nair, J. R., & Gerbaldi, C. (2013). Towards green, efficient and durable quasi-solid dye-sensitized solar cells integrated with a cellulose-based gel-polymer electrolyte optimized by a chemometric DoE approach. RSC Adv., 3, 15993–16001.

48. Song, D., Cho, W., Lee, J. H., & Kang, Y. S. (2014). Toward higher energy conversion efficiency for solid polymer electrolyte dye-sensitized solar cells: ionic conductivity and TiO2 porefilling, J. Phys. Chem. Lett. 5, 1249–1258.

49. Wongcharee, K., Meeyoo, V., & Chavadej, S. (2007). Dye-sensitized solar cell using natural dyes extracted from rosella and blue pea flowers. Solar Energy Materials and Solar Cells, 91, 566–571.

50. Calogero, G., Marco, G. D., Cazzanti, S., Caramori, S., Argazzi, R., Carlo, A. D., et al. (2010). Efficient dye-sensitized solar cells using red turnip and purple wild Sicilian prickly pear fruits. International Journal of Molecular Sciences, 11, 254–267.

51. Zhou, H., Wu, L., Gao, Y., & Ma, T. (2011). Dye-sensitized solar cells using 20 natural dyes as sensitizers. Journal of Photochemistry and Photobiology A: Chemistry, 219, 188–194.

52. Cherepy, N. J., Smestad, G. P., Grätzel, M., & Zhang, J. Z. (1997). Ultrafast electron injection: implications for a photoelectrochemical cell utilizing an anthocyanin dye-sensitized TiO2 nanocrystalline electrode. Journal of Physical Chemistry, 101, 9342–9351.

53. Polo, A. S., & Iha, N. Y. M. (2006). Blue sensitizers for solar cells: natural dyes from Calafate and Jaboticaba. Solar Energy Materials and Solar Cells, 90, 1936–1944.

INDEX

Printed and bound by CPI Group (UK) Ltd, Croydon, CR0 4YY

23/10/2024

01777705-0003